Reordering the World

Reordering the World

Geopolitical Perspectives on the Twenty-First Century

SECOND EDITION

edited by

George J. Demko
DARTMOUTH COLLEGE

William B. Wood
U.S. DEPARTMENT OF STATE

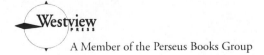
Westview
PRESS

A Member of the Perseus Books Group

William B. Wood is director, Office of the Geographer and Global Issues, United States Department of State. He is the coauthor of the preface and introductory chapters and author of two chapters in the book. The views and opinions expressed there are his and do not necessarily reflect those of the U.S. Government. For the other articles in this book, he served as the coeditor on technical matters. The content of those other chapters is the sole responsibility of their respective authors.

Published in 1999 in the United States of America by Westview Press, 5500 Central Avenue, Boulder, Colorado 80301-2877, and in the United Kingdom by Westview Press, 12 Hid's Copse Road, Cumnor Hill, Oxford OX2 9JJ

Library of Congress Cataloging-in-Publication Data
Reordering the world : geopolitical perspectives on the twenty-first
    century / edited by George J. Demko, William B. Wood. — 2nd ed.
        p.   cm.
    Includes bibliographical references and index.
    ISBN 0-8133-3405-5
    1. Geopolitics.   2. Political geography.   3. International
relations.   I. Demko, George J., 1933–   .   II. Wood, William B.,
1956–   .
JC319.R38   1999
320.1'2—dc21
                                                                98-39355
                                                                    CIP

10   9   8   7   6   5

Contents

v

Tables and Figures

Tables

Figures

Acronyms

APEC	Asia and Pacific Economic Community
ASEAN	Association of Southeast Asian Nations
CIESIN	Consortium for International Earth Science Information Network
CIS	Commonwealth of Independent States
CITES	Convention on International Trade in Endangered Species
CRS	Catholic Relief Services
EC	European Community
EEZ	exclusive economic zone
EU	European Union
FSU	former Soviet Union
GATT	General Agreement on Tariffs and Trade
GIS	geographical information system
GPS	global positioning system
ICJ	International Court of Justice
ICRC	International Committee of the Red Cross
IDPs	internally displaced persons
IGO	intergovernmental organization
IPPF	International Planned Parenthood Federation
MNC	multinational corporation
NAFTA	North American Free Trade Agreement
NASA	National Aeronautic and Space Agency
NATO	North Atlantic Treaty Organization
NGO	nongovernmental organization
NSDI	National Spatial Data Infrastructure
OAU	Organization of African Unity
UNCED	UN Conference on Environment and Development
UNFPA	United Nations Fund for Population Activities
UNHCR	United Nations High Commissioner for Refugees
USAID	U.S. Agency for International Development
WCED	World Commission on Environment and Development
WHO	World Health Organization

Preface

This second edition of *Reordering the World: Geopolitical Perspectives on the Twenty-First Century* is a significant revision of the first but still maintains the original concept of an introductory "political geography reader" covering a broad range of real world issues. Although all of the essays from the first edition that reappear in this one have been updated, several chapters have been replaced based on feedback from teachers and students. We would like to thank all the authors who contributed to either the first or second editions, and especially those who have contributed to both. We were very pleased that the first edition did indeed seem to fill a niche and has been used not only by political geography students but also by those taking courses in international relations and political science. We believe this volume is even better than the first and will attract an even wider readership.

The essays in this volume are not comprehensive reviews of research being done by political geographers, nor are they attempts at defining and explaining all the concepts and theories used in political geography—the former is covered by journal articles and the latter by textbooks. These short essays are more descriptive and issue oriented than theoretical, which reflects a deliberate effort to both attract other social science students to geography and to demonstrate to beginning political geography students that their discipline can contribute to almost every facet of international relations. This reader will have achieved its purpose if even one student is inspired by these essays to pursue further research, which will inevitably yield greater complexity and insight than is provided here.

As with the first edition, this reader begins with the premise that many of today's "global issues" would benefit significantly from a political geographic perspective. Since the first edition was published in 1994, world events have affirmed this notion on an almost daily basis with newspaper headlines that could be taken straight out of the chapters in this volume. The international community faces very difficult challenges as it turns the corner to the next millennium, and political geography will play an increasingly important role in explaining and tackling those challenges. By the end of this book, we believe that the reader not only will agree with this view but also will become one of those dedicated to providing solutions.

William B. Wood
Washington, D.C.

George J. Demko
Hanover, New Hampshire

Part One

Political Geography: Recurring Issues, New Perspectives

Chapter One

Introduction: Political Geography for the Next Millennium

WILLIAM B. WOOD AND
GEORGE J. DEMKO

Political geography is on the ascendance in part because traditional approaches to understanding and tackling a broad range of current problems have failed. The international community is now constantly paying for our weak understanding of the linkages among socioeconomic, environmental, and political forces at local and regional levels—most governments and international organizations still ineffectively react to recurring crises rather than investing in anticipatory analyses and initiating preventive measures. This chronic shortsightedness may be the result of a lingering Cold War legacy that fostered a simplistic world view among diplomats, corporate executives, journalists, and academics—but that excuse becomes less tenable with each passing year. The international community is now well into the "post post–Cold War" period, with leaders around the world claiming to be less focused on various "isms" and more interested in the harsh reality of maintaining viable economies and polities. Political geography, as this volume amply demonstrates, has important insights to offer them.

Geography, of all the disciplines, has most consistently emphasized the fallacy of cookie-cutter social theories and the importance of understanding place- and region-based relationships between societies and the environments they create. These political, economic, cultural, and spatial processes require multidisciplinary approaches if we are to better understand their underlying causes and transnational consequences. Significantly, many of the most pressing political geographic issues are not simply abstract academic concerns. They concern a very fractious world facing the next millennium with such serious and persistent dilemmas as competition among "independent" states within an "interdependent" global economy; widening economic inequities within and among the world's states; widespread human rights abuses; conflicts over natural resource degradation and depletion at local and global scales; tensions and conflicts generated by electoral abuses and ethnic distribution changes; and pressures arising from population growth and distribution. On a more positive note, there are creative roles for new political actors, such as nongovernmental organizations (NGOs), and for innovative new geospatial technologies, particularly for sustainable development applications.

What Is Political Geography?

Although the range of essays in this volume is deliberately broad, the essays are united by a political geographic framework. Political geography is the analysis of how political systems and structures—from the local to international levels—influence and are influenced by the spatial distribution of resources, events, and groups, and by interactions among subnational, national, and international political units across the globe. Such a definition includes group decisionmaking, organization, and implementation activities that affect natural resources. Political geography focuses, on one hand, on how groups interact—particularly the ways they manipulate each other—in the pursuit of controlling resources and, on the other, on how these social, economic, and political activities determine the use of, and thereby modify, the resource base. The resource most often directly implicated in international conflicts is land, whether for intrinsic (it contains minerals or a fresh water source), strategic (it straddles a key trade route), or nationalistic (it embodies a "homeland") reasons. The discipline also assesses the political effects of information and resource flows that change spatial distributions and balances of power.

The political geography of international relations, then, often comes down to control over key resources and flows—be they a commodity such as oil, a specific border crossing, or the "global commons"—and who is best connected in the global system in terms of communications, trade, and idea flows. Apart from territorial and boundary disputes, other less apparent foreign policy issues are also directly influenced by this perspective. For example, the measure of a "regional power" is its exploitative capability, both militarily and economically, over domestic and foreign resources. Many of our humanitarian crises are the result of the tragic inability of different ethnic groups to share resources and the forced displacement of one group by another, setting up flows of refugees and even "brain power" from one polity to another. And, on the global front, much of our concern over population growth, environmental protection, and sustainable development comes down to the fear that our descendants will be living on a smaller, dirtier, and less exploitable region of the earth's surface.

Political geography uses an integrative, regional, and spatial framework that pulls together contributions by both physical and social sciences—it is the one traditional discipline that explicitly bridges the two realms of research. This approach is required to understand and adequately respond to the intertwined trends that underlie most current intrastate crises—often rooted in poverty and income inequality, ethnic intolerance, chronic human rights abuses, and local territorial disputes—which do not lend themselves to grand geopolitical theories. Rather, they require a careful reading and integration of disparate lines of inquiry about local and regional conditions, including long histories marked by conflict and occasional atrocities; modernizing but inequitable economies; dwindling access to arable land; degraded natural resources and polluted environments; territorially complex ethnic distributions; and, perhaps most importantly, the evolving relationships between several thousand ethnic "nations" and the 184 United Nations members that have internationally recognized jurisdiction over the territories they occupy. These nations are groups with a common historical/cultural bond, usually tied to language and/or religion—such as the Sikhs, Miskito Indians, and Kurds—who aspire to internationally recognized sovereignty, or at least political autonomy, for their "homelands." Some states may also be either physically (Bhutan) or politically (N. Korea) isolated, which can impede flows of technology, capital, ideas, and natural resources that make the rest of the global community increasingly interdependent.

A brief review of political geography's role in shaping geopolitical notions of the past century demonstrates that the discipline is not new to debates over foreign policy. Sir Halford Mackinder, member of the British Parliament and political geography's most famous practitioner, revised his famous "heartland" thesis during World War II ("The Round World and the Winning of the Peace," *Foreign Affairs*, vol. 21, July 1943) after first publishing his world regional concept almost a century ago. Although the merits of his arguments about land-based versus maritime-based military powers and the role of the "geographic pivot" in history are beyond the scope of this discussion, he unquestionably played a pioneering role in modern international relations theory. His work is still cited by many political analysts and can serve as a springboard for more refined regional models. Despite his contribution, today's foreign policy makers would be less surprised by the acuity of Mackinder's geopolitical theories than by the fact a geographer was engaged in such thinking at all.

What has kept geographers relatively less involved in the international relations debate for the past half century? Perhaps the biggest single factor is that many post–World War II "human geographers" (those concentrating on socioeconomic and cultural processes as opposed to those studying physical systems) considered "geopolitics" a pejorative term, thanks to another pre-war political geographer, General Karl Haushofer, who played a key role in justifying Nazi Germany's territorial aggression. Indeed, with the notable exception of Mackinder, the roots of modern political geography came from German geographers, with Friedrich Ratzel generally acknowledged as the father of the subdiscipline. His "organic state" theory, now almost a century old, likened the behavior of national governments to that of organisms competing fiercely for resources and lebensraum (living space); ominously, his concept of lebensraum is again being distorted by a new generation of nationalists intent on redrawing political boundaries (for example, the Serbs in Kosovo and Iraq in Kuwait). Thus, although these theories set the intellectual framework for much of this century's "geopolitical" debates, the link between geopolitics and fascism repelled several generations of geographers from the study of political geography at the international level.

The creation of the League of Nations and later the United Nations has apparently done little to dispel the belief that relations among governments are based primarily on distrust and competition. Despite forty-eight years of near universal membership in the UN, numerous wars, fractious alliances, and vast sums spent on military forces seem to confirm that the primary function of a state is to improve its position in the global pecking order or to intimidate potential rivals, with other responsibilities occupying lower priority levels. Although George Kennan's containment theory was directed at the political and economic realms of the Soviet Union, the Pentagon quickly adopted it as a geomilitary strategy because it fit so well into a worldview shaped by the legacies of Ratzel's "organic state," Mackinder's menacing Eurasian "heartland," and Nicholas Spykman's vulnerable "rimland" thesis (another fifty-year-old geographic concept).

After World War II, political geography as a subdiscipline became increasingly irrelevant to most geographers, resulting in few new contributions to international relations debates. Instead, geographers focused on "scientific" research and became intolerant of geopolitical speculations and model building that could not be tested with empirical data and the then fashionable quantitative methods. In the 1980s, political geography was revived, primarily by scholars working within a neo-Marxist paradigm that emphasized spatial patterns and historical processes of inequality at local, national, and global scales. It has been most influenced by social theorists, particularly those exploring the contradictions between an increasingly integrated "world economic system" and the political system of "independent" states. Although this geopolitical systems modeling is insightful, a new generation of political geographers is contributing to international relations through fieldwork-based multidisciplinary research and new tools, such as geographic information systems (GIS) and new spatial models.

Back to Basics

A fundamental lesson of the New World Disorder is that the international community is a much more complex and troubled set of place-anchored peoples than the Cold War and related geopolitical models led us to believe. With the states of the former Soviet bloc caught up in myriad, territorially based economic, political, and ethnic crises, geostrategic containment models lose most of their compelling logic. Rather than the Cold War nightmare of a bilateral military conflict exploding outward toward a global nuclear holocaust, the trend over the past few years has been a geopolitical implosion of internecine atrocities —nation against nation (Bosnia), nation against state (Sri Lanka). Containment is now no longer used to justify preventing the spread of communist ideology but to keep brutal ethnic, national, and territorial conflicts from "spilling over" to neighboring states, as if international boundaries form the rim of a pot holding a bubbling, poisonous brew of hatred.

The imperative for reintroducing political geography into foreign policy after an effective absence of half a century was the collapse of the Cold War. The simple East versus West dichotomy obscured many below-the-surface evils suppressed in the name of empire. Although racism and ethnic intolerance, abject poverty, and environmental contamination were prevalent throughout the Cold War era, their rapid climb up the hierarchy of post–Cold War global problems and their sudden and stark appearance in Europe has confounded the United Nations and its members. Post–Cold War "geopolitical crisis" management seemed firmly under control following Iraq's miscalculated invasion of Kuwait. The U.S.-led coalition's restoration of Kuwaiti sovereignty prompted much speculation on a new era of active UN intervention within, between, and among sovereign states. Just as quickly, however, bold multilateral actions are now being discouraged in light of bogged-down peacekeeping, humanitarian relief, and nation-building efforts in Somalia, Bosnia, Haiti, and Rwanda. Many of these crises have relatively weak immediate national security implications for UN Security Council members but have dramatic consequences on affected regions, the future role of the UN, and the international political order. The failure to quell genocide in Bosnia and Rwanda exemplifies this new multilateral timidity. The saber rattling and international stress over who may have weapons of mass destruction and how to control and monitor them presents another "crisis" for the international community.

Political geography can contribute to this new era of multilateral, multiscale intervention dilemmas by demonstrating how socioeconomic, demographic, political, cultural and environmental factors weave together—or unravel—within and among regions via spatial process and flows and how a negative synergism can lead to a severe humanitarian crisis. Frequent crises, occurring mostly in developing countries, may not lead to nuclear holocaust or send the global economy into a crisis, but they will clearly dominate the UN's agenda and may burst an already strained budget. If political geography is to be of use in grappling with these strategies, it must be applied in an anticipatory manner and in the context of conflict mediation, when noncoercive measures have the best prospect for restoring peace.

Rethinking Regions, Nations, and States

The cliché "think globally, act locally" might be slightly modified by political geographers to read "think and act regionally." A "region" to a geographer is a flexible concept and may encompass any scale and whatever territory is appropriate for a given purpose; depending on the problem at hand, it can be a group of countries or an area within a country. A region, though, must be defined by a common element that ties it together; these ties—that can either pull together or strangle the groups inside it—are at the heart of political geography. The size of a region in this context is important because it determines the parameters of the problem. Regions, whether defined in political, economic, or multiple variable terms, are complex, dynamic, and critical to understanding processes of change. Too often, foreign policy makers seem to perceive only one political unit—the state—and ignore the dynamic political mosaic occurring at subnational scales. A better appreciation of Afghanistan's complex regional geography, for example, may well have led the Kremlin to a different set of intervention policies.

Since the Treaty of Westphalia, the political unit that has dominated international discourse has been the sovereign state, with diplomacy revolving around the groups of states that formed cooperative alliances. The UN is a body politic of member states, who generally recognize each other's legitimacy and sovereignty over a spatially defined territory and population. But it is also a misnomer because it is comprised of political states, not cultural nations, and is far from united on how states should deal with the nations within these states. True nation-states, in which a state's territory contains a predominant, homogeneous cultural group, such as Iceland and to a lesser degree Japan, are few and far between. The aspirations of several thousand nations are not necessarily represented by the member states of the UN. Moreover, the UN has been constrained in its efforts to deal with the nation versus state conflicts because of member state sensitivity to any implied erosion of sovereignty. The inclusion of subnational perspectives, rather than only those of the "legitimate" state, thus may be an important starting point for conflict mediation between an autonomy-aspiring nation and a sovereignty-protecting central government. Such a policy, however, would likely be resisted by those committed to the status quo of centralized political dominance.

Most current humanitarian crises have evolved from sub- and trans-state ethnic tensions and have been greatly influenced by actors other than state governments. On the leading edge of many civil wars are ethnic/national groups whose territory-oriented nationalistic aspirations have been squashed by regimes trying to preserve their traditional, state-defined territorial control. So far, UN members have been lukewarm at best to the suggestion that, in extraordinary crises, the UN has an obligation to override a government's objections and intervene to protect the human rights of an oppressed nation. The UN's intervention dilemma will reemerge time and time again as this fundamental conflict between nations and states erupts into violent civil and cross-border wars. Political geographic reviews of aspiring nations—their relations with "official" state agencies, living conditions, territorial and resource claims, and sense of political and cultural alienation—may help identify future conflicts and perhaps even alleviate those already well advanced. Current examples extend from the Albanians in Serbia's Kosovo province to Tamil demands in Sri Lanka to the Quebecois in Canada, to name only a few.

A regional/spatial perspective within existing states is vital to comprehending such issues as political stability, likely economic tensions and conflicts, and foreign policies. Instability in the former soviet republics, for example, is no surprise to analysts with a thorough knowledge of the regional political geography of the former Soviet Union. The creation of fifteen new independent states has already led to a series of spatial reorientations, with new "core regions" and networks of trade, political influence, and cultural diffusion. The Russian Federation itself confirms the significance of sub-state regions (the 89 oblasts, krays, and national republics) in understanding and predicting the shape of the new government and its range of domestic and foreign policies; the Russian Federal Assembly is largely a reflection of these regional and local demands, biases, and desires for various degrees of autonomy. One of the most important keys to Russia's political stability is the resolution of conflicts between Moscow and its provincial governments.

Political, Economic, and Demographic Boundaries

African insurgencies and massive refugee flows have led to calls for redrawing Africa's boundaries—a debate that strikes at the heart of traditional political geography. These boundaries—imposed a century ago by European colonial powers with little if any indigenous participation—have been maintained by postindependence governments because of a common fear of opening a Pandora's box of ethnic/tribal/territorial conflicts. The argument that once one boundary is redrawn, every tribe in the region would fight to expand or defend its area of control was made by the Organization of African Unity in 1964 and still must be taken very seriously. Although international boundaries, whether in Africa or elsewhere, are no longer held as sacred divisions of the earth, beyond question by those now living within them, they have imposed a geopolitical reality and, once drawn over a territory, are *never* forgotten. Moreover, a new set of boundaries may cause more destruction than resolution.

States use their international boundaries as static barriers over which to negotiate their relations with each other and to define their sovereign space. These boundaries, however, are crude means of preserving the peace among states and do not reflect regional political and economic conditions. Some of their principal functions are already being modified. Simplistic notions of boundary functions ignore—as do many efforts at imposing economic sanctions—the fact that international boundaries are permeable membranes. These political constructs, epitomized by immigration and customs checkpoints, are viewed quite differently by the state institutions that create them and the disenfranchised nations whose homelands are divided by them. Regimes that systematically abuse their citizens hide behind their international boundaries and claim that no multilateral body can impose human rights standards on them within their sovereign space. Although UN and NGO interventions to protect Kurds in northern Iraq challenged the tenet, other, more subtle multilateral agreements—from nonproliferation of nuclear weapons to combating endangered species trafficking—also make inroads across sovereign spaces; but compliance by states is usually voluntary. European Community (EC) integration and the North American Free Trade Agreement (NAFTA) are also voluntary efforts, in essence, to modify the functions of international boundaries by lowering the economic and other barriers built into them. The 200-mile exclusive economic zones in the seas are another example of rethinking political divisions of the earth in an effort to give states more control of their offshore maritime resources.

On the other side of the boundary change question is the specter of more fragmented Bosnias, where a declaration of sovereignty and territorial aspirations unleashed the barbaric practice of ethnic cleansing in numerous towns that had long been tranquil. Boundary and sovereignty changes can cause dormant nationalism to suddenly erupt as the status quo—no matter how unsatisfactory most citizens might believe it to be—is threatened and a safe, prosperous future is equated with one group's control over its perceived homeland. New political leaders can all too easily recall past atrocities and institutionalized discrimination to justify expulsion of "foreigners" or a particular ethnic group from space considered sacred. Once begun, ethnic cleansing becomes very difficult to stop because each displaced group threatens others in the surrounding region and, within the new ethnically homogeneous region itself, the threshold of ethnic purity rises ever upward.

As nations struggle to define their culture, they use both history and geography to justify their right to autonomy and even sovereignty. They map their perceived "homeland"—an exercise made easier with GIS programs—because it helps them transform an ideal goal into a tangible objective. These nation maps, however, threaten the territorial legitimacy of "national" governments and may overlap with other aspiring nations. The boundaries of such maps could lead to internation clashes and even encourage ethnic cleansing and mass population displacement. At the very least, a world political map of nations would be very different and much more volatile than today's version made up of sovereign states and traditionally defined country boundaries. Perhaps the conflict likely engendered by such a map reflects our continuing preoccupation with deterministic histories and ignorance of overlapping geographies.

A nation-based, as opposed to a state-based, worldview would also underscore the tenuous hold that many states have over their territories, with some citizens residing on the periphery of these fragmenting states having much stronger cultural, economic, and even political linkages with their kin in a neighboring state than they do with their own leaders. Some governments have been dealing with "ever-mutating" representations of chaos since their inception and will continue to do so through negotiation and/or coercion. States that hope to ameliorate conflicts within and across their boundaries, however, will have to provide their deeply troubled nations and regions with significant power sharing and access to resources.

Cartographers, like other producers of fiction and nonfiction, depict a world their audience can comprehend, filled with characters—in this case states—with whom they can identify. Politicians manipulate maps as they do any other medium of propaganda. But, then, makers and users of political maps may want to keep changes well ordered because the alternative is a confusing cartographic kaleidoscope. Moreover, governmental influence over map depictions should not be underestimated. Gerrymandering, after all, is a well-established and increasingly accepted international practice, whereby officials stay in power by mapping electoral districts in a way that subverts democratic representation. As the influence of politicians over economic spheres wanes, they may become more vigorous in asserting legitimacy over and manipulating the political spheres depicted on maps; for them, the map's illusion of solidarity is far more pleasant than any regional dissonance outside their windows. In any case, oppression and injustice certainly did not begin nor will end with redrawn boundaries. The key to many global conflicts is learning to share space in a reasonable manner.

Few regions are the sole domain of one ethnic or national group. Several hundred years of mercantilism, colonialism, and imperialism in almost all regions have put in motion many generations of migrants within and among countries. These human flows tend to follow economic opportunities and have been impeded, but not blocked, by increasingly vigorous enforcement of border controls. First-generation immigrants usually congregate together in ethnic enclaves, but subsequent generations—if not impeded by segregationist policies—usually become more interspersed among the indigenous population. Unlike the world at the end of the nineteenth century, when it might have been possible to draw boundaries around many ethnic or national groups, the world in less than one hundred years has become too urbanized, complex, interdependent, and spatially intermixed for simple territorial demarcation based on ethnic or racial homogeneity.

Is it possible to draw new boundary lines when a growing share of the world's population live in historically heterogeneous port cities? How many people will have to be relocated because their presence will no longer be tolerated, explicitly or implicitly, by the new ethnically "pure" state? To where will they move if they do not have a territory of their very own or are themselves products of an ethnically mixed heritage? The most important boundaries, after all, are not those sanctified by the international community but those of the mind. Each new boundary change, then, is a gamble by the international community—an opportunity for a more representative match between a sovereign territory and the people who reside within it versus the all too real likelihood of heightened mistrust and xenophobia and its logical conclusion in forced expulsions and violence toward those who are the new minority. The stakes riding on proposed boundary changes and territorial adjustments are far too high for us to remain ignorant of their underlying spatial patterns and evolving political geographies.

Global Goals—Regional Strategies

Progress toward the global goals of sustainable development and universal adherence to democratic values and human rights will be determined at regional levels and local scales. Sustainable development—the pursuit of economic development objectives while preserving vital resources for use by future generations—is a long-term goal that may remain elusive for most of the world's populations without substantial economic transfers assistance from the developed to developing states. At a minimum, sustainable development efforts must include reducing population growth rates, accommodating intrastate and interstate migration, improving environmental quality, conserving nonrenewable resources, and promoting more efficient energy production and consumption. These laudable goals will require a radical change in the relationship between people and their environments, and governments and their populations, relationships that have intrigued geographers for two thousand years.

Appropriate sustainable development strategies must be tailored to specific local and regional conditions and cannot be designed without an understanding of the political and economic relationships that affect land tenure, resource exploitation, basic needs, and human rights. New geospatial tools, such as GIS and the global positioning system (GPS), and a plethora of remote sensing data are now providing development planners and even local environmental groups with the ability to conceptualize the relationship between "foundation" data ("maps" and satellite images) and a broad range of attribute data (economic, demographic, political, and countless other types of information) within a hierarchy of scales. Such tools, along with an increasing supply of high resolution, satellite-based remote sensing imagery, could empower users to make informed decisions about land cover, use, and tenure, as well as to model the consequences of rapid ecological change, large infrastructural projects, and policies that encourage unsustainable development. They have also become a useful tool in territorial negotiations within and between states, particularly in conflict situations.

Successful sustainable development also hinges on population growth and distribution. Even with greatly intensified efforts at improving the quality and outreach of family planning programs, collectively the population of poor countries will grow by more than 90 million persons per year. The tragedy of this figure, however, will be felt primarily at the local and regional levels, especially where population doubling times are under twenty-five years and economic opportunities are negligible. Rapid population growth in particular regions compounds environmental degradation, economic decline, and political instability. If there are non-Malthusian solutions, they are tied to significant and immediate improvements in literacy, health, and the status of women. Without such changes, "push" pressures on migration will intensify greatly, and hundreds of millions will move to congested cities, the few remaining and rapidly shrinking "resource frontiers," and across international boundaries to places where they are and will be unwelcome.

"Carrying capacity" is an ecological term that has become popularly transferred to humans and food production, but like sustainable development, it too is difficult to measure within interconnected economic systems. Essentially, however, the argument that strained ecosystems are increasingly incapable of supporting the populations that depend on them is substantiated by increasing numbers of "ecomigrants" and increasing dependency on foreign food aid by some countries. The political-geographic dimension of these tragedies lies in analyzing their causes and consequences at various scales. As geographers and other fieldwork-oriented researchers have documented, many current humanitarian crises are not "natural"; they are the results of poor land-use management and the systematic policy of exploitation of rural populations by urban and internationally focused governments. The result of "incapacity" is often widespread malnutrition and even rebellion. Governments, and increasingly NGOs, will have to make very difficult triage decisions. Who should they help, in what way, for how long, and *where?* All too often, the "where" question creates a sovereignty conflict (e.g., Iraq, Sudan, etc.).

The difficult task for international policymakers will be to define how environmental protection and sustainable development can be pursued and carrying capacities enhanced within the constraints and opportunities of a particular region (a state level may be misleading because it is comprised of several regions with very different sustainable development criteria). Debates over natural resource use and the impact of global climate change, for example, are already focusing on regional variations. Efforts to limit greenhouse gas emissions will require analysis of local and regional causes, origins and flows of contaminants, as well as the energy and environmental policies that will affect future production in particular places. Without a regional framework and a realistic measurement of key factors and flows, environmental protection and sustainable development become empty slogans rather than a blueprint for new people-environment relationships.

Similarly, human rights and democracy are more meaningful when seen as a means to ensure that people, especially those who are members of historically oppressed groups, are protected from governmental abuses and have a voice in who represents them at local, state, and international levels. If these goals are to be achieved by the international community, they must be defined in terms of reconciliation between nations and states. Such agreements will be influenced by distributions of ethnic/national groups within states as well as by reasonable policy mechanisms that ensure fairness. For many groups, especially those concentrated in a region bisected by an imposed international boundary, human rights grievances stem from the harsh reaction to their demands for some level of governance over their traditional homelands and resources. Violent conflict over "territorial nationalism" will not ease until significant progress is made on lessening human rights abuses and promoting true representational rule.

Geopolinomics

The magnitude and extent of global economic change renders the old field of geopolitics obsolete. Geopolinomics is perhaps a more appropriate term for the analysis of spatial, political, and economic systems among states and their regions. Geopolinomics focuses on interactions between and among governments and how foreign policies are influenced by regional and global economic prowess, and in turn, how economic linkages and flows are affected by political relations. Rapid shifts among countries in economic wherewithal and enormous global economic flows present dynamic and variegated economic surfaces that must be understood at all scales. The recent economic success of China, for example, has clearly emboldened its behavior in global political circles, but its growing internal, regional inequalities may also condition future foreign policies.

Geopolinomics, then, is the interplay between functional economic and political regions, which are not readily visible on the simple world maps found on the walls of foreign ministries. Yet, such "hidden" regions are often more "real" than those carved by static boundaries because they provide a spatial framework for the most fundamental of transactions, that between producers and consumers. As countries prosper, their economies extend roots—over political boundaries, along convenient transportation routes, and in many ways, across oceans—in ways that absorb or circumvent government-imposed impediments. Regional trade agreements, such as NAFTA, and regional organizations, such as the Association of Southeast Asian Nations (ASEAN), may serve to fertilize regional development, but it is the ever-reaching economic roots that provide the critical nutrients. Although the "gray" regional economy (encompassing both formal and informal markets) may be ignored by multilateral lending institutions because it defies convenient statistical analysis, its force can be felt in any thriving border town.

Infrastructure is the vehicle for most cross-border cultural/economic interactions. Roads, bridges, telecommunication systems, power grids, and railways are the most apparent physical dimensions of these spatial connections. Sometimes a single transportation linkage can serve to bolster the sense of regional integration, as no doubt the sponsors of the recently opened "chunnel" hope for Great Britain and the rest of Western Europe. Even more dramatically, infrastructure can define a new economic region, much as the TVA's electrical grid did more than half a century ago or as the highways and logging roads infiltrating the Amazon do today. Much less tangible than these types of infrastructure are telecommunications linkages. What new types of functional regions are being created, for example, by geostationary satellites that can hook up televisions, telephones, and computers across vast expanses? Some governments have already expressed their concerns over these "cyberspace regions." The system of telephonic, e-mail, and other electronic connections is a network of powerful flows of money, decisions, and information that can, and do, alter power relationships and the globe.

The approval of NAFTA, the trend in Europe toward a more integrated economic union, and the creation of the new Asia and Pacific Economic

Community (APEC) are but a few examples of the reassertion of a new regional geopolinomics. More than half of Asia's trade is intra-Asian and about 60 percent of the European Union's is internal. Trade agreements, such as NAFTA and the reduction of Latin American debt, auger well for a large increase in inter-American trade and investment flows. International, regional, and local trade along with monetary flows over the places of the globe have important implications for local, regional, and global politics and political pecking orders. A peasant revolt in southern Mexico, however, can serve as a sharp reality check that underscores the need to weigh the potential impact of changing hemispheric economic relationships on impoverished and politically alienated groups within deprived regions. Similarly, the failure of trade agreements, the growth of trade deficits, and the erection of trade barriers between groups of states will set in motion other subnational political and economic processes that may be glossed over by state-level statistics.

Flows of capital, trade, services, and ideas within and among states will continue to redefine global patterns of economic influence. In the era of electronic transmission, global financial and monetary flows can change almost instantaneously. Sudden economic declines and weak fiscal and monetary policies, as occurred in East and Southeast Asia in 1997–1998, can significantly affect other regions, states, and locales in terms of interest rates, competition for capital, and political relations. In addition to influences on socioeconomic conditions, sharp and rapid capital flow fluctuations can influence governmental functions, particularly those related to infrastructure, land use, and resource exploitation. Conversely, local and regional governments compete with each other economically by altering regulations and tax structures to better attract capital and/or labor to their jurisdictions.

As we approach the third millennium, unpredictable sets of economic, cultural, and political connections are emerging. New flows of aid to Eastern Europe and the former Soviet Union have been established, while recent crises and international agreements have created new aid targets. Changing cultural and economic flows mark a break from networks dominated by Washington and Moscow and a return to traditional connections with neighboring states; the results are seen in the counter-hegemonic struggles that draw increasingly upon religious and cultural symbols. New economic and energy-driven interdependencies in these regions can readily entrap foreign actors within regional and local disputes.

As with past technological revolutions, "locational" dimensions and interactions change. Modern telecommunications and technology are rapidly altering our measures of space and distance and significantly affecting economic, military, and political relationships around the world. The uneven diffusion of new technologies, in particular, will create new patterns of regional inequity. The ability of a state or region to compete in the global marketplace will largely depend on such translocational technologies as fiber optics, computer networks (particularly Internet-based), and satellite hookups. Ubiquitous microchips can now be designed in Silicon Valley, financed in Zurich, assembled in Malaysia, and sold just about everywhere to fit inside just about every new labor-saving, information-managing device. Even traditional manufacturing centers will be based less on where any natural resource is located and more on where the most cost-effective labor force and most efficient transportation and communication networks are found. International financial systems and networks will transform the "hidden" world map of functional regions by reinforcing connections among once distant places.

Political geography has long been concerned with spatial connections among places and the role of "connectivity" as a surrogate for political and economic relations among regions and polities. The remarkable explosion of near instantaneous communications across the globe has enormous implications for a New World Order. For example, in the 1960s, North America and Europe accounted for over 85 percent of the globe's 150 million telephone lines; now more than 600 million telephone lines and 1.2 billion terminals connect the globe's places and peoples. Direct computer links (e-mail, etc.) and an expanding electronic "information highway" at national and global levels serve as the conduits for an expanding geopolinomic network that will transform the way we think about space and place and will do so at ever-increasing speeds. It may be a small world, but it is enormously complex and dynamic geographically, politically, and economically.

Conclusion

Political geographic analysis is implicit in current international political debates: the viability of multilateral governance; the changing functions of international boundaries; the plethora of nationalist and territorial conflicts; the new technology- and capital-driven flows of ideas and people; and most fundamentally of all, the crisis engendered from rapidly growing, impoverished populations and the degradation of the natural resources on which they depend. But this role is inadequate and too narrow. Perhaps international relations is a more appropriate subset of political geography than vice versa. In an era in which diplomatic formalities may be of declining relevance, it becomes imperative to understand the volatile forces working at nonstate regional scales and to analyze the changes brought about by rapid flows of money, commodities, and ideas.

New types of international connections reach every place up and down the hierarchy of human settlements. Trade and capital flows, migration, and telecommunications now leapfrog over traditional diffusion networks to touch the most isolated village—as long as it has something to offer.

Those places with less and less exploitable resources and more and more underemployed people will be increasingly destitute, anarchic, ignored, and dangerous. Widening gulfs of inequity within and among evolving economic regions and political "homelands" will be altered only slightly, if at all, by current international relations. Although every major international relations decision is stuck to this tattered but resilient geopolinomic web, we have barely begun to seriously think how we might map the strands that bypass governments altogether.

Increasingly, solutions to pressing societal and environmental problems will be assembled from disciplines that do not normally interact with each other, including economics, political science, history, anthropology, strategic studies and—given current environmental and proliferation concerns—such sciences as public health, biology, climatology, agronomy, and physics. Political geography provides an unorthodox but effective framework for bringing together these disparate arenas of knowledge. Demand for its integrative, multiscalar approach will surely grow as decisionmakers working at community, national, and international levels look for new problem-solving strategies. Indeed, the fundamental political geographic issue of how people organize themselves to cope with a rapidly changing environment (natural and human-made) may well be the most pressing concern of the twenty-first century.

Notes

The views presented in this paper are those of the authors and do not necessarily reflect those of the U.S. Government.

Chapter Two

Changing Times, Changing Scales: World Politics and Political Geography Since 1890

COLIN FLINT

Read a newspaper or watch the television news and you are confronted with power: of one country over another as Iraq invades Kuwait, of one ethnic group over another in calculated acts of genocide in Central Africa and the former Yugoslavia, of central government over regions as Sri Lankan forces fight Tamil separatists, and of political parties over the means to redistribute opportunities and wealth. Political geographers have endeavored to understand these diverse struggles for power by looking at both the way space is used by those involved to achieve their goals and how new geographies are created by such struggles. Geography is, therefore, a weapon within political struggles and is a product of their outcomes.

The world of the late twentieth century is much different from that of the late nineteenth century, and the types of power struggles have changed. The methods used to understand these struggles have also evolved with changing times. There are four periods in the history of political geography: from about 1890 to the end of World War II; the 1950s; the 1960s and 1970s; and from the late 1970s to the present.[1] Each of these periods reflects a particular direction within political geography as well as a distinctive period of world history. The close link between the dynamics of world history and the content of political geography is not coincidental.

The first period, 1890–1945, was one of interstate competition as Great Britain sought to maintain its global power in the face of challenges from Germany and the United States. Also, this period was one of dramatic social and political change associated with the increasing power of the urbanized working class and a growth in the ideology of nationalism. The second period, the 1950s, was one of relative prosperity and peace as states rebuilt their economies and societies and many countries experienced the end of colonialism. The third period, 1960 to the mid-1970s, brought an end to the certainties and changes of the postwar period as economies stagnated, development strategies faltered, and the nature of societies was questioned. The fourth period, from the late 1970s to the present, relates to the globalization of production and finance and uncertainties that stem from large-scale flows of investment and goods across state borders.

To help understand how and why the content and attitude of political geography has changed across these four periods, this chapter focuses upon a key analytical concept: geographic scale. Geographic scale is a key component in all political struggles; for example, those in the Civil Rights movement realized that they had to work at the national scale to challenge local and regional injustices. Scale is also a key organizing framework for political geographers trying to understand power struggles. Therefore, it is not surprising that as the dominant political struggles have changed, so has the geographer's use of scale. For each of the four periods, I will outline how scale has been conceptualized within political geography, the societal and intellectual influences on that choice, and the implications of scale on the observations that geographers drew of their changing world.

Geographic scale is the scope of economic, social, and political interactions within different and nested structural layers. For example, certain practices of real estate investment and social legislation can produce neighborhoods dominated by particular ethnic and/or income groups. These neighborhoods are formed at one scale of social activity which, in turn, combine to form another scale of interaction, metropolitan areas. Within a hierarchy of scales, numerous scales can be identified, including, for example, households, firms, regions within nation-states, nation-states, regions of nation-states, and, ultimately, the world economy. No such list of scales is ever complete because geographers use them as units of analysis to frame their inquiries. Therefore, different political geographic studies will identify different sets of scales as being important in answering the particular question in hand.

There are a number of benefits of using scales as the conceptual framework for political geography. First, looking at a variety of scales requires consideration of different political and socioeconomic processes. For example, the Contract with America is a political document that can be understood by looking at economic changes at the global scale and the implications for different social groups at the local scale. Second, scale influences the political actions taken within a particular context. Thus, striking coal miners in Siberia perceive their economic well-being within a place-specific context that influences their choice of political actions and targets. Third, scale allows us to address the interaction between structures and human activity in explaining political actions. The decision to strike is partially determined by nested structures such as the strength of local labor unions and the repressive capabilities of the state balanced against its relative susceptibility to a coal shortage. Finally, use of a variety of geographic scales allows us to see how the multiple identities that people take on influence their political actions. Instead of classifying people solely as members of a particular class, scales force us to see people as parents or siblings, residents of a particular region, and members of a particular race. Political geographers then, use scale to chart multiple influences on political actions, while also noting the particular spatial and temporal context within which these actions take place.

Scale is both a construct of human activity and an academic concept used to study the world. Although scales such as neighborhoods and regions may exist in the world, it does not necessarily mean that political geographers at a particular time will deem them worthy of study. The different political imperatives within each of the four time periods reviewed here produced different societal contexts within which some scales of analysis were studied to the exclusion of others. The history of political geography is a story of the changing emphasis on various geographic scales over time. For each of the four time periods, societal and intellectual influences produced a particular focus, epitomized by key scholarly works, which had implications for both political geography and society as a whole.

Table 2.1 can serve as a guide in our discussion of the four periods. In the first period, competition between great powers was changed by the United Kingdom's declining global influence, giving rise to a political geography interested in the rationalization and justification of interstate conflict. This period witnessed great societal upheavals as the urbanized working class grew and started to make political demands. In response to these perceived threats, conservative scholars published geographic theories aimed at placating domestic political demands through the exercise of power abroad. The conceptual tool was an organic theory of the state that conflated the state and global scales.

In the second period, the dominance of the United States within the global arena precluded concern with international politics. Instead, political geography became increasingly irrelevant as economic growth at home and the rise of a superpower standoff and relative international peace took conflict off of geography's agenda for a while. The result was a political geography that desired to inform the smooth management of domestic poli-

TABLE 2.1 The Historical Evolution of Political Geography

Period	Scales	Key Authors	Intellectual and Societal Influences	Implications
1890–1945	Conflation of state and global scales	Friedrich Ratzel	Neo-Lamarckism	Aggression by states seen as being "natural"
		Sir Halford Mackinder	Great power conflict and domestic changes	Marginalized domestic conflicts
1950s	State	Richard Hartshorne	Assumptions of nineteenth-century social science	Ignored local and global conflicts
			American hegemony and economic growth	
1960s and 1970s	City and global scales	Kevin Cox	Quantitative methods	Brought in discussion of conflict and their causal economic processes
		David Harvey	Marxist social science	
		Andre Gunder Frank	Economic stagnation and challenges to American hegemony	
Late 1970s–1990s	Hierarchy of scales from the individual to the global	Peter Taylor	Immanual Wallerstein	Interlinkages between scales and places
		John Agnew	Anthony Giddens	Economic, social, cultural processes
			Michel Foucault	
			Globalization	Critique of how the world is "written"

cies, using the success of the United States as an uncritical model. The dominant scale of analysis became the nation-state, and the aim was to help oil the cogs of economic growth. A neglect of the global and local scales led to an avoidance of tough questions, such as whether the global economy could accommodate economic growth by all nation-states and also whether all localities would equally benefit from economic growth. In other words, the sources and manifestations of conflict were taken out of the conceptual picture.

The third period in the history of political geography reflects challenges to the role of the United States in ordering the world and a questioning of the ability of the nation-state to provide benefits for all of its inhabitants. Anti-Vietnam War protests and urban riots in the United States and the troubles faced by states made independent through decolonization led to a questioning of the assumptions within a complacent social science. Conflicts and tensions entered the agenda of the political geographer and so did use of scales other than the state level. However, although cities and global economic relations were examined, they were studied in isolation. The implication was that the interlinkages between the scales were not developed.

The fourth and contemporary period is one of globalization. Growing internationalization of states since the 1960s is manifested in unprecedented flows of information, goods, and investment across state borders. As the traditional power of the state is being renegotiated in light of the increasing importance of regional economies and trans-state organizations, a plethora of geographic scales is required for relevant geographic analysis. For example, the environmental sustainability of the global economy, the desire for autonomy or separatism by regions within states, and the awareness of identities other than those imposed by nationalism and Marxism are all problems that characterize the contemporary period. In this societal context of sweeping changes at many scales, political geography has become more eclectic, more critical, and trans-disciplinary. Political geography has returned to a geopolitical perspective but one that sees all political actions, whether undertaken by state forces or social movements, as creating and utilizing control of territory and geographic scales.

Geopolitics: 1890–1945

The first period of political geography began in the 1890s and lasted until the end of World War II. However, some aspects of this early political geography remained in the practice of postwar statecraft. Geopolitics was both an academic and foreign policy enterprise. The goal was a political organization of international space that benefited the ruling elites of a few great powers. Geopolitics involved the classification of spaces across the world political map in terms of their perceived strategic importance to particular governments, who then used this knowledge in their global rivalries. The tool adopted for this "construction" of the globe was an organic theory of society based on a racist social Darwinism and promoted by an antidemocratic elite. The organic theory of the state equated the behavior of states to that of organisms. This entailed two key features. First, states were believed to grow at the expense of weaker states in a competition analogous to the "survival of the fittest." Second, within states different occupational classes were perceived to have particular roles that combined to allow the state to function. Thus, just as the human body needs a brain as well as a spleen, states need agricultural elites as well as street cleaners.

What was happening in the world at the turn of the century that would require or justify the ideology of geopolitics? At the global scale, the expansion of empire and trade had made the globe a closed system. With the imperialist "scramble for Africa" and the opening up of China to foreign influence and trade at the end of the nineteenth century, governments perceived that changes in one part of the world would create changes in another. The interlinkages between people and places was manifest in the growth of international migration and investment. Between 1871 and 1911 about 10.4 million people moved from Britain and Ireland to European colonies. During the same period, the United States received 20.5 million migrants. By the start of World War I, Britain had invested $19.5 billion overseas, which amounted to 43 percent of the world's total foreign investment.[2] Such internationalization signified a new form of economic organization and competition that undermined the role of economic blocs, such as the British Empire. The relative decline of the empire ushered in a period of hegemonic competition as Germany and the United States jockeyed with Great Britain for global supremacy.

With the growing atmosphere of international rivalry came an increase in political radicalism drawn along class and gender lines. Between 1880 and 1914 most Western states were forced to introduce some elements of democracy by giving the vote to women and the working class. Although franchising the working class was limited by techniques that reduced the power of parliament, as in Bismarck's Germany, or by restrictions that made it difficult for eligible voters to register, as in Britain, the result was that the masses had been mobilized into an integral part of politics.[3] This mobilization was facilitated by the increasing urbanization of society. Between 1800 and 1890, Germany's urban population grew from 6 percent to 28 percent. For the same period in England and Wales, the growth was from 20 percent to 62 percent.[4] The final ingredient placed in the domestic cauldron was the injustice of gross income inequalities. The goal of international socialism was to unify a working class diversified by different national, religious, ethnic, and occupational identities. Thus, political relationships were changing on both the domestic and international fronts. The desire of ruling elites to maintain their position at home while competing abroad produced a political geography focused on organizing international space in order to prevent domestic social changes.

The initial interest of political geography was the political organization of space under the rubric of geopolitics. Although most attention has been paid to the aggressive expansionist tendencies of Nazi Germany following the tenets of geopolitical thought, geopolitics came to be seen as an important tool for all countries. In addition, the early geopolitical writers were just as concerned about domestic issues as they were about international events. Success in the international competition for resources was seen as a means of easing domestic troubles and placating the working class. The conceptual tool of geopolitics was an organic theory of the state, which treated the state as a living organism, thereby negating domestic inequities and promoting expansion overseas as a "natural" state goal. An elitist political geography conflated the domestic and international scales through the concept of the state as organism.

Conceptualizing the state as a living organism with "natural" tendencies, rather than a socially constructed institution with strong ties to ruling elites, was made possible by "social Darwinism." Using Darwin's theory of evolution as an analogy, state interaction was seen to follow the logic of the survival of the fittest. Rather than evolution being a random process, as argued by Darwin, political geographers used the ideas of Lamarck to identify social and environmental influences in establishing a hierarchy of power. The result was a scientific justification of racial superiority, which was used to legitimize imperialism. It was this intellectual background that stimulated and supported the academic pursuit of geopolitics.

The view that the state operated as a living organism was expounded by the German geographer Friedrich Ratzel in 1896 and later developed by the Swede Rudolf Kjellén in 1916. The components of an organic theory of the state included the idea that strong states would grow at the expense of weaker states. The strength of a state was a function of its ability to maximize use of its environment. The state as a geographic realm, based on nature-society interaction, became the basis for power. If a society was deemed to be underutilizing its environment, then it could be forcibly evicted by a society better able to exploit it. The genocide of Native Americans by the western expansion of the United States was seen as an example of how this principle worked in practice. If the realm of a state could be effectively expanded, then frontiers were moveable and borders could be redrawn by aggressive states. States on the losing end of such expansion would be deemed to be part of a natural life cycle of birth, growth, and death.

In direct opposition to the aggressive international stance of the organic theory of the state, the view of domestic society was functionalist and nonconflictual. Following a romantic view of an agricultural and feudal order, the organic community was perceived to be

> a community of natural social hierarchies (the male lord and his serfs) bound together by a shared set of mutual obligations and beliefs. Such was a stable community where everyone had a place and knew their duties. The community was made up of a diverse set of people from different social ranks yet all were well-rounded and adept at performing a variety of different practical tasks. Land and soil were at the center of economic life, a life distinguished by honest labor and natural balance.[5]

This idealized view of society was important to the British geographer Sir Halford Mackinder because of the threats he perceived to English power as a result of industrialization and international trade. Writing at the turn of the century, Mackinder believed industrialization and rural to urban migration were destroying the essence of community and, coupled with international trade, preventing his country from feeding itself. If a country was unable to put food on its own table it was home to a decaying community. For Mackinder, the answer was a global political organization of space by a few great powers. This would allow them to share world resources and enable their industries to thrive in supplying imperial markets. The British Empire was, of course, to be one of these units carved onto the globe. Colonized societies were viewed as functional appendages to sustain the well-being of the British empire.

Another concern to the geopoliticians of the time was the expansion of democracy to the working classes. This was seen as a particular threat to foreign policy, which, being "high politics," needed to be insulated from democratic appeals. In a functional organic society, everyone has their role, and no one should be concerned with matters beyond their horizon. Only a few elites had the right to conduct foreign policy, and political geography was their concern.

The theory of the organic society conflated geographic scales. The organic state's natural survival tendency was to expand, and its existence was premised upon a smoothly functioning and nonconflictual society. Domestic and international spaces were seen as complementary though different parts of a natural whole. In a similar fashion, political behavior was not seen to be the result of social processes and conflicts but the expression of "natural" tendencies. This left little room for any discussion of either the global consequences of expanding core states or the possibility of alternative arrangements.

Geopolitics remained a concern for the United States, and other countries, after World War II. As the victorious and hegemonic power, American foreign policy elites required tools for scripting the Cold War conflict with the Soviet Union. However, the United States had also painted a picture of

German political geography as the evil tool by which Hitler was able to plan world domination. Although such tools could be used to inform United States foreign policy, the image of political geography as a tool used for evil purposes negated any popular appeal it might have had in academia. Political geography was, therefore, set for a period of relative inactivity and obscurity.

The Management of the State: The 1950s

After the period of international turmoil and economic uncertainty that marked the first forty-five years of the twentieth century, the United States tried to institutionalize both peace and economic growth. At the end of World War II, the United States had demonstrated its nuclear capabilities and willingness to use them. Using this power, it institutionalized a military exclusion of Soviet influence in Europe through the establishment of the North Atlantic Treaty Organization (NATO) in 1949. In addition, through the Bretton Woods institutions of the International Monetary Fund and the World Bank, the United States institutionalized a liberal global trade and finance regime. In setting up a liberal world order, American ideology assumed that the key social unit was the state. Trade and security formed the postwar geopolitical order.

The assumption of the state's primacy translated into the social sciences, including political geography. Richard Hartshorne's functional notion of political geography emphasized two types of politically organized states: independent sovereign states and dependent countries (colonies, protectorates, possessions, etc.).[6] The latter types of state were effectively ignored by political geographers who focused on the organization of sovereign, that is, Western, states. The state was identified as establishing "complete and exclusive control over internal political relations—in simplest terms, the creation of law and order."[7] New geographic concepts were introduced that would aid in the maintenance of Western states as functional units that unified and controlled their populations.

Hartshorne—the leading American political geographer—excluded several scales from his analysis. First, "lower levels" or regions and localities within states were absent. Indeed, Hartshorne declared that "all the regions of a state . . . have complete loyalty to the overall concepts of the national unit."[8] Obviously, no political geographer could make the same claim today. Second, the political aspects of international trade were given no attention. In an era of free trade managed by the United States, the doctrine of mutually beneficial comparative advantages of free trade went unquestioned. Political geography was used to identify the spatial components that would allow a state to operate, produce, import, and export in a liberal world economy.

The dual intellectual influences of nineteenth-century social science and twentieth-century Marxist/liberal theories created a new focus on "societies" and "development." Societies were uncritically translated into being states, and "development" was the process of transforming agricultural subsistence economies into trading, manufacturing, and consumer economies, following the model of the United States. The influence of Max Weber and Karl Marx created an intellectual framework whereby the state was the obvious unit of analysis and international economic growth on a level playing field was the goal. Political geography's task was to identify potential centrifugal forces that could pull apart those states and propose centripetal forces to keep them glued together.

A functionalist political geography was relegated to a minor role in geography and the social sciences because it divorced itself from analyzing causes of political behavior and the creation of spaces through political action. From an earlier role as a model for international aggression through biological analogies, political geography lost any sense of dynamic process. In a time of prosperity and peace, political geography was simply an intellectual justification of the status quo. Political geographers assumed that since states existed, they should function well, and proper spatial concepts would keep them in good working order. With an exclusive national-level focus, other geographic scales were ignored. The correct spatial tools would integrate regions into an efficient coherent whole, which along with the tenets of free trade, negated any need to look at international economic conflicts.

This period, however, also witnessed the evolution of the Cold War, with its numerous international conflicts and struggles, such as the Korean War (1950–1953) and the Hungarian uprising of 1956. Geopolitics was being adopted by the policymakers of the United States to inform the construction of its liberal world order. Building on the geopolitical analysis and assumptions of Mackinder, Nicholas Spykman proposed a geopolitical vision for U.S. hegemony. In general, international concerns within geography were rejected because of the "geopolitical" legacy of Karl Haushofer as both an academic and policy advisor to Adolph Hitler. Hartshorne's plea that a focus on the functional organization of states was "the central problem of political geography" was an attempt to mark a clear division between the academic study of states and the practice of geopolitics by policymakers.[9]

Analyzing Conflicts and Contradictions: The 1960s and 1970s

The geopolitical world order created by the United States after 1945 and the unquestioned supremacy of the United States did not last forever. In the 1950s, the priming of the Japanese and West European economies by United States dollars led to their economic growth as planned, but this was only possible by forcing the peripheral economies (or Third World) to trade with Europe and the United States. This required a significant financial commitment by the United States, the securing of peripheral regimes against communist influence, and the maintenance of a military presence in Europe to face the forces of the Warsaw Pact. As a result, the American economy was drained in the 1970s as a growing military-industrial economy failed to provide spin-offs for the civilian economy and a growing government acted as a financial burden. These dynamics were to set the context for another change in the fortunes and focus of political geography.

From an upbeat assumption of growth and development came a sense of growing unease. Urban riots in the United States, the Vietnam War, and realization that developmental strategies were not closing the gap between rich and poor countries led to a questioning of Hartshorne's thesis. It became clear that scales within the state were not blindly loyal to the national unit. The use of "state" and "nation" as synonymous terms was now questioned by separatist movements, such as those in Northern Ireland, and by "state's rights" politics in the United States. Persistent inequality between rich and poor countries meanwhile challenged the ideology of comparative advantage in free trade and ushered in a demand for renewed analysis of imperialism.

The response of political geographers to these changing times took two steps. First, geographers analyzed the spatial aspects of problems and proposed spatial solutions. The aim was to ameliorate the operation of the state or world economy. Their next step went further by tracing causes and proposing radical changes to the way society was organized through an incorporation of Marxist ideas into political geography. The dominance of statistical analysis in social science in the 1950s and 1960s impressed a strong quantitative imperative into geography. By the late 1970s, disillusionment with both the narrow questions asked and the quantitative answers obtained by such a positivist approach led to the increasing importance of a Marxist approach that emphasized the economic processes and structures underlying political behavior.

In the late 1960s political geographers began to define themselves in a way that would allow them to engage in contemporary political conflicts and, over time, reinsert political geography as an integral component of social science. In 1969, for example, geographers Roger Kasperson and Julian Minghi defined political geography as "the study of the spatial and areal structures and interactions between political processes and systems" with the intention of focusing on "the interplay between spatial structure on the one hand and process and behavior on the other."[10] Their definition was important because it disavowed Hartshorne's functionalism and introduced the concept of process and a growing awareness of the interactions between societal changes and spatial structures. Such a definition has guided the growth of political geography to the present.

However, the late 1960s was still a period of transition from the functionalist view of political geography to a more critical stance. For example, J.R.V. Prescott's vision of electoral geography downplayed the role of governments in altering the spatial organization of electoral systems to increase their own popularity.[11] Also, electoral geography research did not ask why only rich industrialized countries had liberal democracies. Elections were, according to Prescott, simply a means of organizing states. On the other hand, a few geographers were beginning to analyze "the voting decision in a spatial context" by identifying how neighborhood characteristics influenced individual political decisions.[12] Thus, in the subfield of electoral geography, both the functionalist and analytical approaches were vibrant.

The subfield of border studies offers another illustration of an evolving political geography during the 1960s. Border studies had acted as a bridge between geopolitics and functionalism as it was used as a tool by states negotiating boundary changes after World War I. In the 1960s, the functionalist approach studied the effects of borders on frontier regions and interstate relationships, such as trade and migration flows. New computer-based spatial analysis tools improved capabilities to measure and map interactions across borders, but there was no investigation of what the borders themselves were defining—the state, which remained as an unquestioned and obvious institution.

In the 1970s, political geographers began to look at societal problems using a more systematic spatial-analytic approach.[13] The spatial-analytic approach adopted the dominant quantitative techniques and concepts in the social sciences but added distance and location. For example, Kevin Cox looked at the locational aspects of individual decisionmaking within the city.[14] In particular, he studied "externality effects," which modeled the impact of one actor's choice on other actors, such as the impact of the pollution from a factory on the surrounding population. Cox's approach introduced a much-needed analytical approach to political geography by incorporating rational choice models into the spatial aspects of urban conflicts. As opposed to Hartshorne's assumptions, Cox acknowledged inequities and conflicts within states. However, as in Hartshorne's work, only one scale was studied—in this case, the city—thereby ignoring the political role of the state as a filter between the world economy and the city.

The spatial-analytical approach to political geography was highly significant because it introduced analytical rigor and moved the subdiscipline toward the identification of societal problems. What was lacking, though, was a discussion of what social processes were causing those problems. Cox's models assumed that individuals had the freedom to act rationally, which precluded a discussion of how political power was organized in cities. These urban geographies concentrated on individual firms and households within the city, rather than on influences coming from outside the city, and thus limited the identification of social processes shaping space. Cox's work nonetheless acted as a catalyst, and soon political geographers were using a Marxist framework to identify economic structures underlying inequalities in income and life opportunities.

The transition in political geography from a traditional approach that helped justify and serve capitalism to a Marxist rejection of the status quo was manifest within the pages of one particular book. David Harvey's *Social Justice and the City* began with a liberal stance but concluded that such tinkering with the system was inadequate and that a radical upheaval of society was necessary.[15] Harvey later went on to provide a geographic interpretation of Marx in *The Limits to Capital*,[16] arguably one of the most important works in modern political geography. It argued that space or geography is the product of economic processes; capital created places through investment and labor, but new investment opportunities required that these places be continually destroyed and replaced by new ones. Economic tension also existed between the desire to benefit from new investment opportunities and the need to secure return from existing ones. In other words, old landscapes could not be torn down to make new ones until they had been productive for a while. Thus, capital created a particular landscape that mediated the next round of investment decisions. Harvey thus introduced the notion that space is both a product of social activity and a mediator of political behavior.

At a different level, the emergence of many new independent states freed from the shackles of colonization and their subsequent difficulties in raising living standards introduced a new mode of thinking about the global scale. The "dependency school," a sociological approach, proposed a core-periphery structure in the world economy and argued that these poor states could not "develop" by merely copying what Western countries had done.[17] Political geographers were influenced by the dependency school's thesis of inherent inequity in international relations. In other words, a global economic structure existed that restricted possible development options of states. Just as Cox had decided that scales within the state needed to be studied, the dependency school realized that relations between states also required analysis.

Challenges faced by the United States to its ordering of the world revived academic engagement of geopolitics. Saul Cohen's "Geography and Politics in a World Divided" was intended to inform U.S. foreign policy by introducing the geographical concept of the region.[18] By adapting geopolitical concepts from Mackinder, Cohen identified "shatterbelts," strategic regions the United States should target in its conflict with the Soviet Union. The coexistence of the dependency school and Cohen's geopolitics illustrates the varied perspectives during this period of political geography. While the dependency school was challenging the intellectual framework that the

United States used to order the world, Cohen was identifying ways for the United States to adapt its foreign policy to the challenges it was facing. Thus, one of these approaches was a direct challenge to the world order while the other was proposing means of maintaining that order.

An important similarity between the works of Cox, Harvey, and the dependency school is their respective foci on one particular scale, even though it is a different scale from the state. Interactions between scales are not mentioned, as if processes occurring at one scale are isolated from those at other scales. At the end of the 1970s, that paradigm changed, allowing for the consideration of processes other than economic ones and also relaxing the deterministic tendency of some of the Marxist works.

The End of American Hegemony and the Linkage of Geographic Scales: Contemporary Political Geography

The 1970s marked a period of international and national turmoil. The end of the Vietnam War in 1975 marked a decline in the ability of the United States to impose its security order on the world. The oil crisis of the previous two years had transferred significant wealth and with it political power from industrialized countries to oil producers. Unemployment and inflation rose in the industrialized countries, and a sense of crisis in Europe was promoted by terrorist attacks in Germany, Italy, and Great Britain. To cap it all, punk rock groups proclaimed that Western society really was culturally bankrupt and major changes were on the horizon. However, after a brief period of more "liberal" government, the elections of Ronald Reagan in the United States and Margaret Thatcher in Great Britain helped to settle their respective establishments, and renewed economic growth followed after some harsh economic adjustments. What the 1970s turmoil brought home was an awareness of the interconnectivity of the globe and the interlinkages of local to global scales. Unemployment and inflation at home, for example, were related to the global recession and U.S. foreign policy through the oil crisis and the value of the dollar.

Three important intellectual influences on the revitalization of political geography in the late 1970s can be identified. First, the combined efforts of the dependency school and Immanuel Wallerstein's world-systems analysis introduced the idea of structural and historic constraints on a global economy.[19] States acting in a particular way were constrained by a world economic structure that required a few rich core states and a greater amount of poor and peripheral states. Talk of "development" was cheap as the structural necessities of the world economy would prevent most peripheral countries from entering the "developed" core.

Second, Anthony Gidden's interpretation of structuration theory loosened up Marxist determinism, including Wallerstein's.[20] In Gidden's theory, societal and global structures do constrain the actions of agents (individual actors and institutions), but agents still possess some freedom to change the nature of the structure. For example, the geopolitical order of the Cold War was a structure that constrained the actions of states, but states were still able to act within it. Some states, such as Yugoslavia, enjoyed a relatively large amount of freedom of action despite the security regime that the Soviet Union tried to impose across Eastern Europe. Over time, the actions of the United States, the Soviet Union, and other states led to the collapse of the Cold War order and the current reordering of the world. Embracing structuration theory, geographers looked at how a variety of structures influence the actions of agents at different scales (households, regions, or political parties, for example).

Third, postmodernism, as envisioned by the French philosopher Michel Foucault, introduced the notions of discourse and the social construction of space.[21] Through words and actions people construct spaces as part of political struggles. The creation of barricaded streets and "no-go" areas by rioters in Los Angeles or Lagos, for example, are a means of challenging state power by creating a space of resistance. Thus, the role of individuals in creating space is reemphasized. Similarly, the use of language in describing the world has political implications. For example, the use of the term "periphery" instead of "Third World" is a discourse that argues that poor countries are part of a global power structure rather than separate units who have yet to develop. Conscious of this responsibility, geographers began to acknowledge their own perspective and discourse within the field of political geography.

The combination of domestic concerns and a global perspective has returned political geography to an agenda that at first glance is similar to that of traditional geopolitics. As Mackinder and the German geopoliticians were aware, domestic concerns were related to international processes. But contemporary political geography is also quite different from the earlier geopolitics. First, the "organic state" processes that "naturalized" military aggression have been totally discredited and replaced by a less deterministic model of human agency operating within the context of particular structural constraints. Second, political geography has become critical of the way geopolitics has been used to "objectively" push the interests of a particular state. Third, national security is no longer perceived as simply a military concern but as an economic and cultural program. Finally, the state is no longer the unquestionable dominant actor or only level of analysis. Without prematurely declaring the death of the nation-state, the power of the state with regard to transnational and subnational scales has been renegotiated.

The similarity between the political geography that emerged at the end of the 1970s and beginning of the 1980s and the old school of geopolitics lies in their mutual interest in the political organization of space rather than its functionalist and spatial-analytical aspects. Instead of state elites trying to organize international politics to their benefit, the politics of the world economy occurs at a variety of scales, and many more actors are trying to recreate the world political map. The dynamics of politics in a changing world and a frustration with the legacies of geopolitics and the functionalist approach helped reestablish political geography as an integral part of current social science.

Geographic scale was used as a new organizing principle at a conference of political geographers in 1983, designed to show the subdiscipline as a viable part of the International Geographic Union.[22] Geographic scales were seen as social constructs used to the advantage of some actors and to the disadvantage of others. For example, phases of British-Irish politics produced different politics depending upon the scale of division. Before 1921, the entire British Isles constituted a British/Protestant/Unionist majority whereas the current two states provide an Irish/Catholic majority and a British/Unionist majority. The scope of this conflict shifted through the creation of new political entities that formed different political outcomes. In other words, scales are not just given but are created, and there is an important political element to this construction.[23]

The papers of the 1983 conference ranged from studies of local government through regions and the state to processes of international conflict. Scales were no longer treated exclusively and interlinkages between scales were emphasized. Peter Taylor's framework, for example, identifies three key scales.[24] Starting at the locality (or place) level, classified as the "scale of experience," one experiences political and economic dynamics emanating from the global scale, or the world economy—the "scale of reality." In between these two scales is the state, the "scale of ideology," which seeks to ameliorate the impact of the global "reality" on the individual's experience. For example, the processes of postwar great power competition at the global scale between the United States and the Soviet Union created employment opportunities within particular American localities related to the defense industry. The state secured and dispersed the necessary funds to finance this employment, whether it was at an army barracks or in the manufacturing of tanks and aircraft carriers. As the competition between the two powers waned at the global scale, so did the need for defense-related jobs, turning the local experience into layoffs rather than jobs. Again, the state intervened by protecting some jobs or, at the least, paying benefits to the unemployed. Taylor calls the state "the scale of ideology" because its political actions deflect attention away from the global scale where the dynamics are really taking place, the world economy. Therefore, politics remains fragmented into many different state movements and parties rather than unified into global movements.

The emphasis on geographic scale created a contextual approach to politics. John Agnew and Doreen Massey emphasized the importance of interactions between the specific context of places and economic and political processes operating at a different scale.[25] The spatial distribution of political processes—such as revolutions and coups, social movements, voting behavior, local government politics, separatist nationalism, and famine—can be examined using this contextual approach. Place-specific institutions, such as the church or trade unions, can help people form a particular mental picture of the world, which they can use to respond to problems arising at the local, state and global scales. Context is, then, a function of the nature of a place and its interlinkages with the world; it can help explain the patterns of political behavior and place-specific variations and similarities. For example, the context of the core and periphery structure of the world economy can help explain the distribution of coups d'état in poorer countries, but each coup will be shaped by unique institutions and histories.

In combination, geographic scales and context illustrate the way that space mediates political actions. By defining the scope of a conflict and the role of place and other spatial constructs in mediating political behavior, geography becomes an essential component of political practice. As captured in the title of Agnew's seminal work, *Place and Politics,* the social construction of geography and politics are intertwined.[26] For example, Adolph Hitler's rise to power was the result of a process that required the construction of a political movement, the Nazi party, and a related geography of power that created electoral regions of supporters. However, contextual influences, such as a history of anti-Semitism in some regions and a tradition of working-class loyalty to left-wing parties in other regions, created a pattern of relatively high and low Nazi party support.[27] Hitler tried to reduce the strength of this regional resistance by phrasing the scope of important issues at the level of a unified German nation rather than as local class issues. Hitler's message to the German people thus resonated through the dynamics of global recession and great power competition.

An understanding of the global scale has been a necessary component of political geography since its revival. The global economic recession of the 1970s and early 1980s, the Third World debt crisis, and the question of the relative decline of the United States combined to provoke a consideration of the global context and its impact upon politics at other scales. Taylor's adoption of Wallerstein's world-systems analysis saw global economic relations as determined by a core-periphery relationship while great power competition and economic restructuring both followed a broad cyclical pattern.[28] Wallerstein's theory has been influential, even on his critics, in its view that economics and politics are inseparable. The term "geo-economics" captures the spatial practices and implications of the ordering of the world economy.[29] The expansion of the United States' military reach over the globe and the U.S. attempt to regulate the world economy are different tactics in the same strategy of creating a liberal world order aimed at benefiting the American economy. However, the creation of one global order can unintentionally create another; thus the globalization of U.S. manufacturing and finance undermined its relative position.

The world-systems and the geo-economic approaches both differ from the earlier geopolitics in their refusal to draw a picture of the world to inform the actions of a particular state. Instead, theoretical concepts are used to explain and understand the dynamics of the world. Another consequence of contemporary societal uncertainties has been to challenge the way the world is represented by geographers. Political geographers, who are usually white, Western, and male, order the world as they wish it to be represented in their work. Geography becomes "geo-graphy," the writing of the world. Their particular representations are thus biased and other geographic representations need to be considered.[30]

This latest challenge, called critical geopolitics, has helped political geographers distance themselves from traditional geopolitics by exposing the biases and intentions of the statesmen who wrote it. Contemporary political geographers have also critiqued their own worldview as the product of a particular stance. Critical geopolitics also challenges the use of particular words by politicians who use them to sell their worldview to the public. For example, President Bush's use of the term "quagmire" when referring to Bosnia recalled images of Vietnam and, therefore, was intended to reduce calls for American intervention. On the other hand, proponents of American intervention to stop the genocide in Bosnia used the term "holocaust" in a tactic intended to recall images of Nazi death camps. In other words, part of the political conflict over U.S. policy was the deliberate use of different representations of "Bosnia" to the public. Contested images became part of the political struggle over the definition of U.S. foreign policy.[31]

A revitalized and contemporary political geography is widening geopolitics in two ways. First, it no longer refers to political/military models of the world but incorporates economic and cultural processes at scales other than the global. The overthrow of apartheid's restrictive social and spatial structures can be classified as a geopolitical struggle just as readily as the Gulf War. Second, geopolitics is not the preserve of elites but a tool of various competing social institutions and groups; the rioters and the police, the Mexican army and the people of Chiapas. In direct contrast to traditional geopolitics, contemporary political geography exposes a range of structures and discourses of power instead of trying to create them as part of statecraft.

Political geography is now establishing itself around an organizing framework of interacting scales. This allows geography to encompass political issues at any scale. But instead of conflating scales or prioritizing one particular scale, relationships and tensions between scales provide the incentive for geographic inquiry. Political geographers seek to understand the interactions between scales and not, as Mackinder tried to do, negate them.

Conclusion

These are exciting times for political geography not only because of changes in the world but also because we view them differently. Political geography allows for the discussion of political events at any scale, whether it be civil war or trade war, an event in your home town or across the globe. Political geographers are excited by discovering new possibilities of political action within particular geopolitical and geoeconomic constraints and opportunities. What are the possibilities for peace, for example, in Northern Ireland, given the constitutional changes proposed by the newly elected Labour Party? What are the constraints on Russian politicians in light of NATO expansion and domestic economic woes? Are the chances for the destruction of chemical weapons better or worse in the post–Cold War world?

These are just a few of the challenging issues we face in the immediate future. As we debate policies that affect our future—locally and globally—we need to keep in mind how geographic scale can be used to define the scope of a conflict and make local to global connections transparent. Understanding how the language of power can define particular places and spaces in foreign and domestic politics will enable critical thinking about political choices.

Notes

1. Although political geography existed in various forms prior to 1890, this chapter focuses on the past one hundred years or so.

2. These figures are taken from Gearóid Ó Tuathail, "Putting Mackinder in His Place: Material Transformations and Myth," *Political Geography* 11 (January 1992): 100–118.

3. See E. J. Hobsbawm, *The Age of Empire, 1875–1914* (New York: Sphere Books, 1989), pp. 84–111.

4. Hobsbawm, *The Age of Empire, 1875–1914*, p. 343.

5. Ó Tuathail, "Putting Mackinder in His Place," p. 109.

6. Richard Hartshorne, "The Functional Approach in Political Geography," in *The Structure of Political Geography*, ed. Roger E. Kasperson and Julian V. Minghi (Chicago: Aldine Publishing, 1969), pp. 34–49; reproduced from the *Annals of the Association of American Geographers* 40, 1950.

7. Hartshorne, "The Functional Approach in Political Geography," p. 35.

8. Ibid., p. 43.

9. Ibid., p. 35.

10. Roger E. Kasperson and Julian V. Minghi, eds., *The Structure of Political Geography* (Chicago: Aldine Publishing, 1969), pp. xi, xii.

11. J.R.V. Prescott, "Electoral Studies in Political Geography," in *The Structure of Political Geography,* ed. Roger E. Kasperson and Julian V. Minghi (Chicago, Aldine Publishing, 1969), pp. 376–383.

12. Kevin R. Cox, "The Voting Decision in a Spatial Context," *Progress in Geography* 1 (1969): 81–117.

13. In the 1970s, the studies that were to become the foundation of contemporary political geography were usually written by geographers who had established themselves in other subfields (such as urban or economic geography) but became engaged with political issues and processes.

14. Kevin R. Cox, *Conflict, Power, and Politics in the City: A Geographic View* (New York: McGraw-Hill, 1973).

15. David Harvey, *Social Justice and the City* (London: Arnold, 1973).

16. David Harvey, *The Limits to Capital* (Oxford: Basil Blackwell, 1982).

17. See A. G. Frank, "Sociology of Underdevelopment and the Underdevelopment of Sociology," *Catalyst* 3 (1967): 20–73; and A. G. Frank, *Dependent Accumulation and Underdevelopment* (New York: Monthly Review Press, 1978).

18. Saul Cohen, *Geography and Politics in a World Divided,* 2nd ed. (New York: Oxford University Press, 1973).

19. See Immanuel Wallerstein, *The Capitalist World-Economy* (Cambridge: Cambridge University Press, 1979).

20. Anthony Giddens, *The Constitution of Society: Outline of the Theory of Structuration* (Cambridge: Polity Press, 1984).

21. For a discussion of the influence of postmodernism on geography see Derek Gregory, *Geographical Imaginations* (Cambridge, Mass.: Basil Blackwell, 1994).

22. Selected papers from this conference were published in Peter Taylor and John House, eds., *Political Geography: Recent Advances and Future Directions* (Totowa, N.J.: Barnes and Noble Books, 1984).

23. This example is taken from Peter J. Taylor, "Introduction: Geographical Scale and Political Geography," in *Political Geography: Recent Advances and Future Directions,* ed. Peter Taylor and John House (Totowa, N.J.: Barnes and Noble Books, 1984), pp. 1–7.

24. See Peter J. Taylor, *Political Geography: World-Economy, Nation-State, and Locality,* 3rd ed. (Harlow, UK: Longman Scientific and Technical, 1993), pp. 40–48.

25. John Agnew, *Place and Politics* (London: Allen and Unwin, 1987); and Doreen Massey, *Spatial Divisions of Labour* (London: Macmillan, 1984).

26. Agnew, *Place and Politics.*

27. Colin Flint, *The Political Geography of Nazism: The Spatial Diffusion of the Nazi Party Vote in Weimar Germany* (Unpublished Ph.D. diss., Department of Geography, University of Colorado, 1995).

28. Taylor, *Political Geography: World-Economy, Nation-State, and Locality.*

29. For an application of "geo-economics" see John Agnew and Stuart Corbridge, *Mastering Space: Hegemony, Territory and International Political Economy* (New York: Routledge, 1995).

30. See Gearóid Ó Tuathail, *Critical Geopolitics* (Minneapolis: University of Minnesota Press, 1996).

31. This example is taken from Gearóid Ó Tuathail, *Critical Geopolitics,* pp. 187–224.

References

Agnew, John. *Place and Politics.* London: Allen and Unwin, 1987.

Cox, Kevin R. *Conflict, Power, and Politics in the City: A Geographic View.* New York: McGraw-Hill, 1973.

Hartshorne, Richard. "The Functional Approach in Political Geography." In *The Structure of Political Geography,* edited by Roger E. Kasperson and Julian V. Minghi. Chicago: Aldine Publishing, 1969, pp. 34–49; reproduced from the *Annals of the Association of American Geographers* 40, 1950.

Harvey, David. *The Limits to Capital.* Oxford: Basil Blackwell, 1982.

Mackinder, Sir Halford. "The Geographical Pivot of History." *Geographical Journal* 6 (1904): 421–437.

Ó Tuathail, Gearóid. *Critical Geopolitics.* Minneapolis: University of Minnesota Press, 1996.

Ratzel, Friedrich. "The Territorial Growth of States." *Scottish Geographical Magazine* 12 (1896): 351–361.

Taylor, Peter J. *Political Geography: World-Economy, Nation-State, and Locality,* 3rd ed. Harlow, UK: Longman Scientific and Technical, 1993.

Chapter Three

Geopolitics in the New World Era: A New Perspective on an Old Discipline

SAUL B. COHEN

The post–World War II era has come to its decisive end. With its bipolar global geopolitical system now only historic memory, scholars and statesmen are searching for new paradigms to anticipate the direction that the international system might take as it enters the twenty-first century. A "new" geopolitics—one offering fresh perspectives on the relationship between geography and politics—is important to the development of realistic paradigms and the spatial conceptual basis for the new world map.

The outlines and stability of this new map depend on equilibristic forces. Many students and practitioners of world affairs viewed the end of the Cold War, Germany's 1990 reunification, the collapse of the Soviet empire, China's opening to the world, and the United States–led coalition victory over Iraq in 1991 as harbingers of a New World Order. Such hopes were quickly dispelled. Almost immediately after these stunning events, bloody nationality and religious conflicts broke out in the former Soviet Union (FSU) and Yugoslavia, and Saddam Hussein, still in power in 1998, ruthlessly quashed Kurdish and Shia uprisings. He has also faced down the threat of U.S. military retaliation against his undermining of UNSCOM's operations. Rebellion in Congo has toppled its long-serving dictatorial President Mobutu, Albania's authoritarian "free-market" ruler was ousted, Kosovo is in upheaval, and the Arab-Israeli peace negotiations have stalled. Expectations of a New World Order have turned to fears of a world in chaos.

The world is neither in order nor in chaos, but experiencing the isostatic disturbances and short-term disequilibrium of a global geopolitical system entering a new stage that will be maintained by perturbations and contradictions. The new world map now unfolding reflects this dynamic balance.

Perturbations that affect geopolitical equilibrium are constant and span the globe. Contradictions abound and often cause the perturbations. While economic transnationalism weakens international barriers, resurgent ethnic nationalism sparks savage local wars and uproots minority populations. Ultimately, the excesses of war and population displacement come to an end through a combination of exhaustion of warring factions and international intervention: witness the end of factional conflict in Lebanon, the disintegration of the Khmer Rouge in Cambodia, the Dayton-imposed peace in Bosnia, and the winding down of civil wars in Liberia and Sierra Leone.

As socialist structures have crumbled and the market economy has begun to emerge, a Russia rich in resources and scientific and technological manpower is overwhelmed by the rise of corrupt and criminal new elites, the erosion of governmental fiscal capacities, food and energy shortages, and continued rural poverty. In time, however, foreign capital and technology will help break the resource logjam, and a united, rejuvenated modernized Russia will become a surplus producer. This same Russia, freed of repressive centralized communist rule, has veered toward chaos and the threat of dismemberment as autonomy-seeking groups and independent state monopolies seek to go their own way. But the pendulum is likely to swing back toward more centralization as the threat of atomization and dedevelopment sinks in.

Countries struggling with drought, famine, disease, fratricidal warfare, and capital investment shortages are now of lesser interest to the major powers that once valued them for the strategic military bases they provided. The resultant diminished flow of foreign assistance has in some areas increased human suffering and political unrest. When such disturbances reach the flash point of endangering global markets, international assistance programs for less developed countries will likely be reinvigorated (although foreign aid has often been diverted by corrupt rulers rather than targeted to development objectives). A dialectic process thus operates to balance the swings of a geopolitical pendulum.

Geopolitics was very much the basis for the Cold War, the divisions of Germany and Korea, the wars in Southeast Asia, Afghanistan, and the Middle East, and the string of superpower military bases that gird the world. This was the Cold War era, influenced by the "old" geopolitics whose theoretical basis was a crude nationalistic spatial determinism framed within the broader context of the Eurasian Continental and Maritime realm competition for supremacy.

The "new" geopolitics could foster an era of accommodation based on the recognition of a spatial and political interplay among localism, nationalism, and internationalism. A useful theory for this "new" geopolitics is one that conceives of the political earth as a unified system evolving in developmental stages (Cohen 1991).

"Old" and "New" Geopolitics

A short definition of geopolitics is "the applied study of the relationship of geographical space to politics." Geopolitics is therefore concerned with spatial patterns, features, and structure, and political ideas, institutions, and transactions. The territorial frameworks of such interrelationships vary in scale, function, range, and hierarchical level—from the national, international, and continental to the provincial and local. The interaction of spatial and political processes at all of these levels creates and molds the international geopolitical system.

Political geographers have a responsibility and opportunity in these extraordinary times to bring geopolitical perspectives that are balanced and objective, something that the "old" geopolitics was not. Indeed the leading Soviet and American geopoliticians who designed the Cold War strategy were not geographers and had only limited understandings of geography.

Except for its rejection of racist superiority theories, Cold War geopolitics drew much from the environmental and organic determination of German Geopolitik. The chief exponent of Geopolitik, Karl von Haushofer, provided the geographical rationale for Nazi world conquest (Whittlesey 1942). The ultimate goal of German geopoliticians and their Nazi masters was world dominance. Germany's seizure of the Eurasian heartland, the geographical pivot land of the world—a concept adopted by Haushofer from Halford Mackinder (1904, 1919)—would assure Germany command of Mackinder's "World Island" (Eurasia and Africa), and thereby the world. Although regional relations were central to Haushofer's doctrines (as they were to those earlier geographers on whom he drew), his spatial frameworks were unidimensional and failed to account for differing hierarchical levels of territorial frameworks.

In American geopolitics, geography was simplified and distorted to serve political ends. The American geopoliticians, many of whom were European émigrés, came either from the scholarly fields of international relations and history (e.g., Strauz-Hupe, Walsh, Kennan, Kissinger, Brzezinski) or from the military, not from the ranks of academic geography. For these geopoliticians, geography generally means distance, size, shape, and physical features all viewed as static phenomena. The idea of geography as spatial patterns and relations that reflect dynamic physical and human processes is absent.

The "old" geopolitics appealed to its American practitioners because it simplified the world map. The world could be divided into two without taking into account underlying and changing subdivisions. Pseudo-scholarly American geopolitics produced the doctrine of rigid containment and the "falling domino theory" that influenced so much of the foreign policy thinking set forth by John Foster Dulles in the 1950s. It also led to rejection of strategies to accommodate overlapping spheres of influence (such as in the Middle East) or Cold War neutrality (e.g., India, which upon achieving independence sought to maintain neutrality).

American geopolitics helped plunge the global system into nearly half of a century of military buildups, arms transfers, and regional and local conflict. Although the strategy did lead ultimately to the bankruptcy of the former Soviet Union and speeded the dissolution of its empire, it also led the United States into a destructive and strategically flawed war in Vietnam and into an arms race that sapped the American economy and weakened its social fabric. The remarkable economic strides of the 1990s, characterized by low inflation and low unemployment, reduction of national deficits, and increased information-age productivity have occurred in a post–Cold War era in which the proportion of United States GNP dedicated to defense spending has decreased to 4 percent and trade access to the entire world has increased.

The rulers of the FSU also espoused a crude geopolitical doctrine—Eurasian expansionism. In the interest of geographical security, Moscow seized and annexed vast areas in Eastern Europe, expelling millions of indigenous peoples from border areas and dispersing countless numbers of ethnic minority groups within its territory, while promoting the movement of Russians eastward. It manipulated changes in the boundaries of its "satellite" states, controlled their space militarily, and monopolized their resources and markets. Territorial expansion and oversight were considered necessary to promote the security of the Soviet motherland and to safeguard the Marxist revolution.

In contrast to the "old" geopolitics, which was an instrument of war, the "new" geopolitics can be applied to the advancement of international cooperation and peace. In an era of global accommodation, there will continue to be local wars, both formal and guerrilla, especially as underlying ethnic, religious, and cultural separatist forces generate proliferation of nationalized states. The "new" geopolitics focuses on the evolution of this political world as an interdependent system at varying scales—from the national and transnational to the local. It calls attention to those political areas, either independent states or parts of sovereign entities, that play special locational roles in linking the international system. Moreover, it embraces the geopolitical study of transnational economic, social, and political forces, as well as cities, and the impact of changing technology on population movements.

By analyzing the interdependence of economic, cultural, social, and political process within changing spatial milieus, the new geopolitics can influence military strategic considerations and place economic competition on the same plane as military competition. For example, it challenges the Western premise that the "southern" continents (Subsaharan Africa and Latin America) should be consigned to marginality because the need for Western military bases there has been so dramatically reduced. This region remains globally important in a variety of other ways: as a source of environmental warming, pollution, drugs, and certain diseases, and a base for global terrorism; as a world market with dynamic growth possibilities but also as a potentially destabilizing element in world credit and commodity markets; as a generator of regional and global migration streams; and as a locus for massive starvation, genocidal conflict, and human rights abuse. Most recently, vast offshore oil fields in the Gulf of Guinea have attracted large-scale international petroleum company investments in this part of West Africa—a region, until recently, ignored because of its seemingly hopeless poverty and corruption. This politically and economically unstable region can thus become a catalyst for global instability. It warrants sustained strategic and economic attention even though the need for Cold War military bases has passed.

Dynamic Equilibrium and Change

Depending on one's worldview, the global map reflects a system in order or disorder, in balance or imbalance. Recipes for creating order on such a map are hardly new. Immanuel Kant's "universal international state" called for a universal mechanism to enforce the peace as early as 1795.

The United Nations and the League of Nations before it were attempts to fashion world order based on partnerships of national interests. The League failed and the United Nations has often been bypassed. The power of veto held by the five permanent members of the Security Council has often undermined the UN's ability to take decisive action.

Transcending this new associative world system, the two superpowers created a bipolar world of strategic realms and regional subdivisions that coexisted through a tenuous balance of shifting forces. This was not "order" as a strongly regulated or fixed arrangement of ranks and clusters but a Cold War balance based on a dynamic equilibrium. This type of worldwide order has persisted and even been enhanced by the end of the Cold War.

This was an equilibrium in which disturbances abounded. Not only did both the United States and the USSR engage directly in major wars, the former in Korea and Vietnam and the latter in Afghanistan, but they supported surrogate conflicts in many parts of the world.

The boundaries of the world's geostrategic realms and geopolitical regions were in flux during this era. Geopolitical regions or parts thereof changed their political and economic orientations, and nations switched alliances. The balance was maintained by superpower fear of mutual nuclear destruction, limiting each from overreaching in the global competition, and by the rise of other first- and second-order powers with independent interests and agendas. This international equilibrium was neither a static nor a closed system. Rather it was equilibrium created by balancing opposing influences and forces in an increasingly open system.

In general, balance is maintained not only by what Adam Smith referred to as an "invisible hand," or the rational self-interest of peoples. In the absence of reason, excesses of war, economic greed, and environmental imbalances ultimately encounter resistance. When things go too far, there is reaction, correction, regulation.

As geopolitical systems mature, their parts multiply and draw power away from the center. As they decentralize, territorial units acquire increasing self-responsibility, and interactions become self-directing. This interaction may be either competitive or cooperative, but it is almost always turbulent. Without such turbulence there is no change, and without change there is no progress.

A major manifestation of such change is the reorientation and realignment of political territorial units. Regrouping occurs at all levels of the geopolitical scale and is not spatially random or independent of lines provided by nature. Instead its geopolitical cleavages occur along specific fault lines which are drawn from an array of optional boundaries provided by nature (Cohen 1973). The relative strength of particular cores determines where and at what hierarchical scales geopolitical repartitioning takes place.

Evolution of a General Organismic System

Treating the world as a general organismic system provides insight into the relationships between political structures and their operational environments. These interactions produce the geopolitical forces that shape the system, upset it, and then lead it toward new levels of equilibrium. To understand the system's evolution, it is useful to apply a developmental approach. Such an approach is derived from theories advanced in sociology, biology, and psychology. The developmental principle holds that systems evolve in predictably structured ways, that they are open to outside forces, and that hierarchy, regulation, and entropy are important characteristics.

Combining organismic concepts from Herbert Spencer's mid–nineteenth-century sociology works with those of Heinz Werner (1948), a psychologist, and Ludwig von Bertalanffy (1966), a psycho-biologist, provides the foundations for a spatially structured geopolitical theory. It is a theory that is holistic, concerned with order and process of interconnecting parts, and applies at all levels of the political territorial hierarchy.

To adapt the developmental principle geopolitically, I hypothesize a system that progresses spatially in stages. The earliest is undifferentiated. Here none of the territorial parts are interconnected, and their functions are identical. The next stage is differentiation, when parts have distinguishable characteristics but are still isolated. The highest stages are specialization and hierarchical integration. Specialized and complementary outputs of unique territorial parts leads to system integration, while hierarchical structure directs output flows.

World War II and the end of colonialism paved the way for new world geopolitical arrangements. It is thus a logical starting point for tracing the development of the current system. In the early postwar years, the two bipolar realms controlled by the Soviet Union and the United States were clearly differentiated from each other. Within each realm, except for the superpowers, the parts were relatively undifferentiated. This was the time when nations had begun to recover from the ravages of World War II. There was dominance, but only a one-step hierarchy within either realm. Both Stalin and Dulles believed that superpowers should dominate all parts of their respective geostrategic arenas without the need or interference of intermediaries.

That system quickly changed. Specialized regional cores such as Common Market Europe and Japan arose, initially as junior partners and now as friendly competitors to the United States. The Warsaw Pact also emerged as a geopolitical and economic bloc but could not shake off Soviet control until the Soviet Union collapsed. Within the Eurasian realm, China soon challenged the USSR for strategic parity. These new power centers began to develop independent ties to other states and regions.

A number of regionally important states began to emerge in the 1970s, giving added substance to the regional structure. These second-order power states tried to impose a hierarchy of their own within their respective regions. India became dominant in South Asia, defeating Pakistan in war and casting its stamp over the rest of South Asia. Nigeria, not the United States, has led peacemaking efforts in Liberia and Sierra Leone conflicts and is now in a position to exercise regional military control over the oil-rich coastal waters of the Gulf of Guinea. Vietnam, with help from the Soviet Union, drove the Khmer Rouge from power in Cambodia and has become a dominant force in the rest of former Indochina, although China still looms as the transcendent power in the region.

Although hierarchy remains a major structural element of the world system, it does not follow a rigid rank order. States can exert influence over others without always having to defer to those in the rank above them.

Thus, Albania broke away from Tito's control to reach directly to the USSR before splitting with its Soviet masters and turning to China. Mexico and Venezuela defied the United States and tried to shape an independent Central American policy in bringing peace to Nicaragua and Guatemala.

Flexibility in the worldwide hierarchy reflects both the maturation of individual states and the fact that geopolitical power relations are no longer a function of distance alone. Air, sea, and telecommunications allow ties to develop between states that are relatively far apart. Flexibility is further enhanced by the impact on individual states of transnational corporations and international social and political organizations.

This increasingly complex and open world system can be described as a "polyocracy." The system has overlapping spheres of influence, varying degrees of hegemony and hierarchy, national components and transnational influences, interdependencies, and pockets of self-containment. It is all the more complex because its parts are at different stages of development. Regional states thus play differing roles according to their spatial and economic interactions with major powers and neighbors.

Entropy

Helping to link the international system is the drive of the less mature parts to rise to levels already achieved by more mature sectors. The balance of relationships across and within the nested regional frameworks can be analyzed in terms of entropic and hierarchical conditions. This provides some guidelines to help determine levels of development.

A key element in the dynamism of the system lies in shifts in power among different states and regions. Some power changes are the result of domestic developments, involving political reorganization, new economic structures, or changing social patterns. Others can be attributed to external national and transnational forces. Three orders of national power—the first or major, the second or regional, and the third or subregional—affect the balance of the global system, but even lesser-order states are change-agents influencing regional and global patterns—witness Angola, Afghanistan, and Ethiopia.

Entropy level is indicative of a state's or region's relative capacities to influence events within various orders of power and is also a useful measure of balance in geopolitical relations. Defined in physical systems as "the availability of energy to do work," entropy always increases as stored energy becomes exhausted. Thus without energy infusions, a system's ability to work constantly declines. If the world were to consist of closed geopolitical units, then each unit would ultimately collapse.

Only hermetically sealed systems, however, behave according to this law of inevitability. Geopolitical entities whose leadership seeks to close them off from outside forces do suffer from reduction of energy by the exhaustion of their human and natural resources and resulting high levels of entropy; ultimately, however, human needs and strivings pry open these "closed" systems. They become recharged through the introduction of peoples, goods, and ideas as "free energy." In particularly favored open systems with high energy inflows, such as the United States or Singapore, entropy can be negative; they have increased their abilities to "do work," and thus their citizens have enjoyed rapidly rising standards of living. In contrast, the Soviet Union, Albania, and Myanmar experienced dramatic increases in entropy levels as a result of their decades-long and ruinous attempts to close their systems.

Criteria that can be used to measure entropy include savings rates; agricultural yields; manufacturing productivity; debt repayment; percentage of R&D exports; numbers of patents, scientists, engineers, and foreign scientific exchanges; and fuel-use efficiency. In general, regions fall into four categories: low entropy, consisting of the U.S.-Canadian core of North and Middle America, the European core of Maritime Europe and the Maghreb, and much of Offshore Asia; medium entropy, consisting of the Heartland, Eastern Europe, the Middle East, and East Asia; high entropy, consisting of South Asia; and very high entropy, consisting of Subsaharan Africa and South America.

Levels of State Power

The Global Geopolitical Structure

The "geo" in geopolitical analysis starts with spatial structure. To understand geopolitical systems, we must address the spatial categories that geographers use as frameworks of analysis. The structure is hierarchical. At the highest level are two geostrategic realms: the Maritime and the Eurasian Continental (see Figure 3.1). Below the realm is the geopolitical region. Realms are arenas of strategic place and movement. Their trade orientations differ, the Maritime being open to specialized exchange, whereas the Continental is inner-oriented. Regions are historically shaped by contiguity, migrations, and political, cultural, military, and economic interaction.

The Maritime realm has a global reach. Within it are geopolitical regions that constitute the second-level geopolitical of the hierarchy, including North and Middle America, Maritime Europe and the Maghreb, Offshore Asia, South America, and Subsaharan Africa. The Eurasian Continental realm consists of two geopolitical regions: the Soviet Heartland and East Asia.

Most of the second-level regions are contained within the realms. Three, however, lie outside. South Asia is an independent region. The Middle East

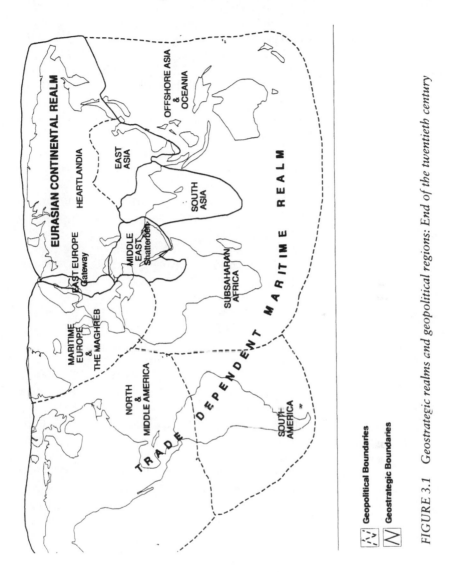

EURASIAN CONTINENTAL REALM

HEARTLANDIA

EAST EUROPE
Gateway

MARITIME
EUROPE
&
THE MAGHREB

MIDDLE
EAST
Shatterbelt

EAST
ASIA

SOUTH
ASIA

OFFSHORE ASIA
&
OCEANIA

SUBSAHARAN
AFRICA

NORTH
&
MIDDLE AMERICA

SOUTH
AMERICA

TRADE DEPENDENT MARITIME REALM

Geopolitical Boundaries

Geostrategic Boundaries

FIGURE 3.1 Geostrategic realms and geopolitical regions: End of the twentieth century

is a Shatterbelt, a zone of contention caught between the two realms. The third, Central and Eastern Europe, could be an emerging Gateway region. This transitional zone has the potential to facilitate contact and interchange between the two realms, unless the West tries to follow up NATO's recent expansion (the Czech Republic, Hungary, and Poland) with the inclusion of additional East European states—a step highly likely to draw a new "Iron Curtain" between the Maritime and Eurasian Continental worlds.

The third level of the hierarchy is the national state. States are hierarchically ordered according to their power positions and functions in the world system. Gateway territories are a special category. Currently many Gateways are components of the subnational, or fourth level, of the hierarchy. These are embryonic states that can accelerate regional or even global economic exchanges; they can stimulate the evolution of larger nations, from which they have spun off, or can achieve considerable economic autonomy, which better enables them to fulfill their exchange functions (i.e., Hong Kong).

The world system is in a continuing process of development, as it moves toward greater specialization and integration. As national energies and transnational forces gain or lose momentum, the regional frameworks—realms, regions, states, and subnational units—change in status and in boundaries. This, in effect, produces new relationships within the system, which requires continual rebalancing.

Despite the profound changes that have taken place in the world in recent years, the basic framework of two geostrategic realms remains—the Trade-Dependent Maritime World and the Eurasian Continental World. Of the world's five major power centers, only one is now both a military and an economic colossus: the United States. Two, Russia and China, are formidable military forces. Despite the collapse of the Russian army and navy, it retains its nuclear forces. China has a powerful army and growing navy, but its massive rural population is mired in poverty despite the industrial economy's phenomenal growth. Two power centers, the EU and Japan have global-dominating economies but have not mobilized militarily in ways that are equal to their economic capacities. Because each lacks vast strategic space and is vulnerable to the military pressures of near Chinese and Russian neighbors, their strategic alliances with the United States remain their strongest security cards.

The deteriorating economic and political fortunes of the former Soviet Union may suggest that the concept of a Eurasian geostrategic realm no longer has validity. However, it is premature to dismiss Russia from its dominant perch over the Eurasian Heartland. A revived, albeit smaller and loosely confederated union, such as the Commonwealth of Independent States (CIS) under Russian leadership and ideologically compatible with its East European neighbors, will remain in a position to dominate its geostrategic realm. This realm is characterized by a distinct set of interrelationships expressed in terms of patterns of circulation, economic orientation, and historic, cultural, and political traditions.

In defining realms as Continental and Maritime, the reference is not only to lands and climates; it also describes worldviews. The Eurasian Continental realm is more isolated, more inwardly oriented, and more heavily endowed with raw materials than its Maritime counterpart. Its people have deep ties to the land. Whatever happens to the CIS there will be a Russia and some allied or subordinate republics to occupy the Eurasian Heartland. "Heartlandia" will remain a large, well-endowed, and technologically advanced power, capable of influencing events in much of the rest of the world.

China, too, belongs to the Continental realm; it is not part of the Maritime World as portrayed by Mackinder and Spykman in their times and Richard Nixon in his. The vast majority of Chinese live off the land, not from sea trade. Even with China's recent spurt in commerce and its growing trade surplus with the United States, it accounts for only about 2.5 percent of the world's trade in goods and services—compared to approximately 18 percent for the United States. For those living in riverine North and Central China, the mountain and the river hold mystical attraction. Only in coastal Southeast China does the Maritime tradition prevail.

Russia's and China's common border strategically links them so that they cannot turn their backs on one another; they have to find a modus vivendi. Even though political change in the former Soviet Union is in sharp contrast to China's quashing of political democracy stirrings, the Chinese resistance to political change must inevitably give way, especially as its commitment to a market economy, now enhanced by its regaining of Hong Kong, intensifies. When both Continental powers no longer are trapped by competing versions of Marxist ideology and enjoy more open systems, they are likely to find more in common.

The two anchors of Offshore Asia are Japan and Australia. Forty years ago the prevailing view was that Australia was geopolitically aligned with Great Britain and the United States; today it is clear that Australia's destiny is more like that of any Asian-Pacific nation. Australia sends 60 percent of its exports to Offshore Asia, and 22 percent of its total merchandise trade is with Japan, its leading trade partner. In addition, 40 percent of its annual immigrants are Asian. As an exporter of raw materials, Australia's main challenge is to raise its high-cost manufactures to competitive levels with its neighbors who are also part of the globe-encircling Maritime realm.

Japan's situation as the dominant economic and political power in Offshore Asia is unique because of its reluctance to exercise military and economic pressures, although it has such a large stake through its capital and trade investments there. This reluctance manifests itself in the way in which Japan has approached the financial collapse of Offshore Asian countries during the 1998–1999 crises. Although it provides much of the rescue capital, it has permitted the International Monetary Fund and the United States to take the lead in pressing for fiscal reform measures in those countries. This is the reverse of South Asia, where India freely applies military options, or East Asia, where China has been militarily involved in both Korea and Indochina.

South Asia belongs to neither geostrategic realm. It has unique geopolitical regional status. In their early history, especially from its Indus Valley beginnings in 3000 B.C. to Roman times, the peoples of India were seafarers and colonizers. In ensuing centuries they became more continentally oriented, although South Asia was an important source for special raw materials and a market for imported goods during British rule. As an independent geopolitical region dominated by India, South Asia remains rural-based and largely continental, with a reach that has not expanded widely across the Indian Ocean. This does not minimize the growing importance to the region of overseas trade, shipping, and modern-day immigration. However, the basic orientation is inward (India's merchandise trade is only one-fifth that of China's)—a condition that explains the limited impact of extraregional contacts upon the geopolitical objectives of South Asian states.

If the intensity of trade interactions were the only criterion for defining major geopolitical regions, then South America and Subsaharan Africa would not qualify. However, the combination of their unique physical and cultural regional characteristics both sets them aside from and yet connects them with other parts of the Maritime realm, especially the United States and Europe. Subregions in both South America and Subsaharan Africa are more distinct political, military, and economic arenas than are their larger continental regions.

In forging the geopolitical unity of the South American continent, the weight of the eastern Brazil–La Plata core plays an overwhelming role. Moreover, Chile's strategic relations with Argentina, the vulnerability of the Central Andean countries' rainforests and savannas to Brazil's demands, and Colombia's ties with Venezuela will inhibit western South America from gaining independent geopolitical status on a par with the east.

Subsaharan Africa's subdivisions—Southern Africa, Western and Central Africa, and East Africa—are arenas of far more intense political, cultural and military interaction than is the continent as a whole. When the two strongest regional powers, Nigeria and South Africa, sort out their internal problems, they may carve out two distinct geopolitical regions, subsuming the smaller and weaker central and eastern subdivisions. This would create two geopolitical regions—the South and East African Lands of the Indian and South Atlantic, and the West and Central African Lands of the Mid-Atlantic.

The much-touted world system does not really span the entire globe. Perhaps it never will. Parts of the world, especially in Subsaharan Africa and South America are outside the modern economic system and do not benefit from the exchanges that are so important to the development process. The merchandise trade of these two regions is only 3 percent of world exchange. With the exception of pockets of modernity in Brazil, Argentina, and Chile, and in South Africa, these regions are relatively untouched by the capital flows, technology transfer, and specialization of industry that characterize the developed market economies, which generate over 70 percent of world trade. The continents centering around the South Atlantic, but also bordering the Pacific or Indian Oceans, represent the quarter of the world's land and ocean areas that have been referred to as the "Quartersphere of Strategic Marginality." Mired in poverty and plagued by corruption, their economies suffer from chronic overproduction of commercial crops, minerals and their substitutes, and world competition. Only the prospect of vast oil finds in the waters off West Africa offers a ray of economic hope for Subsaharan Africa.

President Clinton's trip to Subsaharan Africa in 1998 was heralded as initiating an effort by the United States to increase trade and to stimulate capital investment in Africa. Whether this highly publicized journey meets its objectives or remains a public relations footnote to history remains to be seen.

Although dominated by the United States and European Union power centers, the "Quartersphere" is marginal in a strategic sense. Naval and air strike forces, long-range air weapons, and satellite surveillance capabilities have minimized the significance to the Maritime World of southern continental land bases. Pipelines and Suez now account for as much oil movement as the shipping routes around the Cape of Good Hope. The Panama Canal now takes most of the Pacific shipping trade oriented to the United States East Coast.

Unless the West's military, strategic, and economic disinterest in the two regions is subsumed by humanitarian considerations and the imperative of spreading democratic processes and establishing political stability, small-scale bloody conflicts are likely to continue there, sapping the vitality of the developing world and discouraging much needed outside investment and assistance.

Shatterbelts

The concept of the Shatterbelt has long been of interest to geographers, who also have used the terms "Crush Zone" or "Shatter Zone." Mahan (1900), Fairgrieve (1915), and Hartshorne (1944) contributed pioneering studies of such regions.

A Shatterbelt is a politically fragmented area of competition between the Maritime and Continental realms. By the end of the 1940s, two such atomized regions had emerged: the Middle East and Southeast Asia. They were not geographically coincident with previous "Shatter Zones" because the global locus of geostrategic competition had shifted. The former East and Central Europe Shatterbelt had fallen within the Soviet strategic orbit, and the Maritime and Continental Worlds became divided by a sharp boundary in Korea.

Kelly (1986) has noted that other parts of the world are also characterized by conflict and atomization usually attributed to Shatterbelts. Although wars, revolts, and coups are chronic in Central America and the Caribbean, South America, and South Asia, the distinguishing feature of the Shatterbelt is that it presents a playing field used by two or more competing major powers from different geostrategic realms.

Thus South Asia is not a Shatterbelt. India's dominance within a divided South Asia is not seriously threatened by the United States, Russia, or China, despite Chinese victories over India's armed forces in the border wars of 1959 and 1962. The Central America and Caribbean subregion is under America's military strategic and tactical sway and was so even during the period of Soviet penetration into Cuba.

Shatterbelt regions and their boundaries are fluid. During the 1970s and 1980s, Subsaharan Africa became a Shatterbelt. The Soviet Union used its Cuban surrogate as well as its Eastern European satellites to provide military and technical support to Ethiopia, Angola, Namibia, and Mozambique. Its adjoining Middle Eastern bases were important jumping-off points for Subsaharan Africa. The then USSR also made political inroads into Guinea, Mali, Congo, and Tanzania. With the collapse of the Soviet Union, the region shifted back to the Maritime realm.

Another major change in the geopolitical map is that Southeast Asia has also lost its Shatterbelt status. Its insular and southern peninsular portions have become economically and politically part of Offshore Asia and the Maritime World. Malaysia and Thailand now enjoy considerable industrial development, with their economies linked to Japan and the United States. This has followed Singapore's remarkable growth as part of the Maritime realm and the realignment of Indonesia with the West and its Offshore Asian neighbors.

Now that Russian military and economic forces have retreated from Southeast Asia, Indochina will soon likely fall back within the East Asian sphere. Vietnam will have to find a strategic accommodation with China. An isolated and impoverished Myanmar (Burma) will be left, with almost no foreign trade or contacts. When the military regime is eventually overthrown and the country opens itself to the world, it will become more oriented to Southeast Asia, suggested by its 1997 admission to ASEAN.

The only remaining Shatterbelt is the Middle East but it, too, is in transition. Since Russia has ceased to be a major economic and military supplier, the region has tilted toward the Maritime realm. Russia remains sensitive to the future strategic orientation of the new Caucasus and Central Asian states, especially to the roles of Turkey and Iran in the futures of its former Asian republics. It cannot be expected to remain quiescent if an anti-Russian Muslim coalition arises in the northern Middle East. However, the era of broad Heartlandic penetration throughout the region, with bases in the Red and Arabian Seas and in the Eastern Mediterranean, is over, at least for the near future.

In the post–Gulf War world, the European Union is likely to exert more influence over the Middle East and to emerge as the second major intrusive power. European influence over Iraq, Iran, and Turkey is on the increase as that of the United States decreases. The lead taken by Britain and the EU in calling for a "safe haven" for Kurds in Northern Iraq and later for the Shi-ites in the south are examples of this renewed influence, as is France's active role in mediating the 1997–1998 crisis in Iraq.

Besides outside intrusions, the Middle East is also a Shatterbelt because it is so highly fragmented internally. Alliances among its various states are fluid and serve as forces for destabilization. The United States and others can help in the quest for regional stability, particularly by pressing for elim-ination of weapons of mass destruction, reduction of conventional arms, and commitments to act against new regional aggressors. It can also redou-ble its efforts to mediate the Arab-Israeli conflict. But outside powers can-not guarantee against continued turbulence. The challenge is to contain re-gional tensions and to minimize their impacts since it is not likely that they can soon be eliminated.

Collapse of the Soviet Union has set the possibility for the Middle East's geopolitical transformation from a Shatterbelt, whose western half may be-come affixed to Maritime Europe creating a new "Euro-Mediterranea," while its eastern half operates as a small independent geopolitical entity or merges with lands to its north or east. With Russia no longer a major re-gional intervenor, the West is relatively free to seek solutions that will bring a measure of unity to the Mediterranean parts of the region.

This prospective geopolitical region could become a reincarnation of the Roman Mediterranean-Atlantic world, with boundaries extended into Cen-tral Europe (see Figure 3.2). Euro-Mediterranea would coalesce around two sprawling industrial/postindustrial cores—the Channel/Rhine and the Gulf of Valencia/Gulf of Genoa axes. Creation of the region depends, how-ever, on Turkey remaining oriented to the West and NATO, and resolution of the Arab-Israeli conflict.

If the Middle East were to divide geopolitically, the states from the Tigris-Euphrates through the Iranian Plateau might join with what is left of Afghanistan, Pakistan, and perhaps Turkmenistan and Tajikistan to form a separate Islamic bloc. Iran's role as a strong regional power would be piv-otal if it could resolve political aspects of its religious differences with its Sunni neighbors. Were such an Islamic bloc to emerge, the Eastern Arabian peninsula, and particularly Iraq, would become a small Shatterzone.

Continuity of Major Core Areas

Continuity, as well as change, characterizes the present and previous geopo-litical systems. Major power cores—the Atlantic/Great Lakes United States,

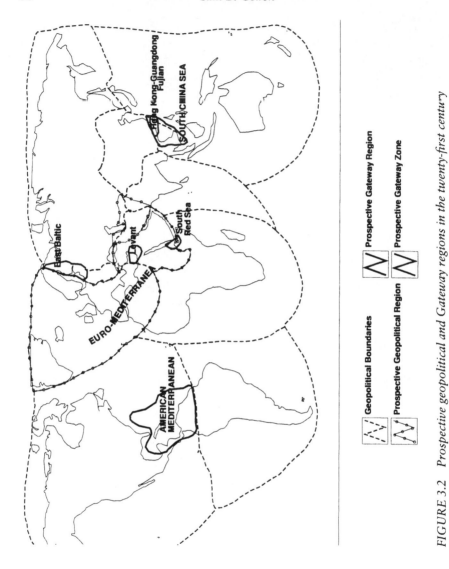

FIGURE 3.2 Prospective geopolitical and Gateway regions in the twenty-first century

the Russian Heartlandic industrial/agricultural triangle, the Western European/North Sea industrialized arc, the riverine plains of Northeast and Central China, and the Central and Southern Japan Inland Sea conurbations—persist as the system's major geopolitical cores, although secondary ones have emerged in California and the adjoining Southwest, and on the South China coast (see Figure 3.3). These cores maintain their economic and political importance in the face of ideological and economic upheavals and the geopolitical reorientation of their peripheries. The growth and interaction of

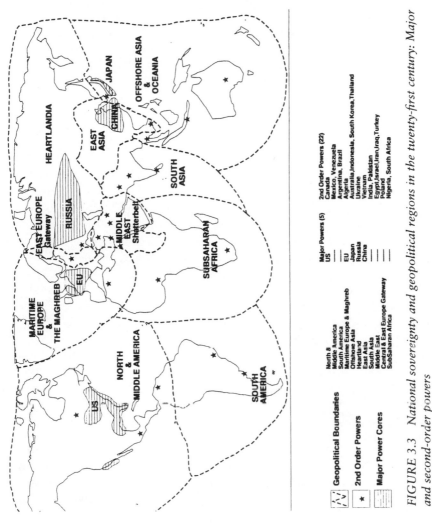

FIGURE 3.3 *National sovereignty and geopolitical regions in the twenty-first century: Major and second-order powers*

these five major power centers can be a significant force for stability and continuity in the system, as it has been a cause for conflict in the past.

The strength of these power cores is a function of their human and physical resource bases and their locational settings, as well as the political energies generated by overarching historically held ideals. The Maritime United States, North Sea Europe, and Inland Sea Japan mobilize their total state frameworks around global exchange functions. Russian "Heartlandia" and Riverine China are land-oriented in their bases and settings. Their strategic focus is on continental interior peripheries; maritime contact, although of increasing importance, does not radically change the base.

The test of the permanence of a major geopolitical core is its ability to withstand loss of substantial parts of its periphery. Even without the Soviet empire, and even if Ukraine should fail to develop a meaningful federative link with it, the Russian core persists as a world power. With its 125 million people, this industrialized triangle, extending from St. Petersburg to Moscow to Rostov on the Don and then eastward across the Urals to the Kuznetsk Basin, also has fertile agricultural lands, rich deposits of oil and gas, iron ore and metallic minerals, and a vast pool of scientific and technical manpower. With its vast resources, Russia is positioned to overcome the travails of its transition from a totalitarian communist state to an open, democratic, and market-oriented system.

Riverine core China, too, can withstand loss of part of its periphery. This core is contained in the river valleys of the Yellow and Yangtse and their coastal extensions, flanked by the grassy plains to the northwest and Manchuria. In contrast to the coastal South, the highly mineralized and heavily industrialized Chinese continental core with a population of over half a billion is culturally isolated and economically self-contained.

Should the south China coastal zone centering around Kwantung and Hong Kong unexpectedly split off, Riverine China would still be an independent colossus. The China that is now opening itself to outside investment and the world exchange economy with such astonishing rapidity is the Maritime China of the Southeast, not the northern and central core. If South China were to break away in a major upheaval, or if Tibet or Sinkiang were to do so, the northern and central core would still be a world power center.

The American core once consisted only of the United States manufacturing belt of the Northeast and Great Lakes, backed by the Midwest prairies agricultural belt and southern Ontario; now California and its southeastern extension into Arizona has emerged as a second core of world-class standing.

Should California and the remainder of the United States–Canadian Pacific coastland ever break away politically to seek greater geopolitical orientation with the rest of the Pacific Rim—a highly improbable scenario to say the least—the Atlantic/Great Lakes core, with its 115 million people and unparalleled financial, manufacturing, research, and agricultural resources would still remain a power center of world-class proportions.

Western and North Sea Europe, the EU's economic heart, is also a core area of global magnitude. What makes North Sea Europe a persistent world power core are its skilled population of 175 million and the industries, markets, and agricultural and mineral resources of the belt that extends from midland and southeast England through the north of France, Benelux, Germany, and northern Italy. Were the United Kingdom to reject European unity over a common currency, the damage to the core would be comparable to the geopolitical orientation of Ukraine away from Russia toward Eastern Europe. In both cases the power of the cores would be seriously but not fatally damaged.

Because its core is so geographically compact and therefore inseparable and its national periphery is of such limited importance, Inland Sea Japan is unique. Extending from the central and southern plains of the eastern Honshu coast (Tokyo-Yokohama, Osaka-Kobe, and Nagoya) to northern Kuyushu, this tightly packed coastal region of 80 million people contains the bulk of Japan's manpower, economy, and infrastructure. Japan has experienced considerable territorial expansion in modern times. Korea, Manchuria, and Taiwan were conquered and then lost in a span of half a century. However, the long-term impact on the core was negligible. Japan has gained through international trade what its outlying territories or former empire could never have provided economically.

Second- and Third-Order Powers

Regional or second-order powers are emerging cores within their regions (see Figure 3.3). They have nodal characteristics in terms of trade and transportation, and military influence, and they aspire to regional or subregional influence. Limited extraregional economic or political ties are also characteristic of such powers. Finally, although often overshadowed by a major power, second-order states try to avoid satellite status, sometimes by playing off one major power against the other.

Although second-order states may have regional hegemonical aspirations, their goals are constrained by geopolitical realities. Where such states are located within the same region as is a major power, they cannot mount a serious challenge. Elsewhere, even though major powers are absent, proximate and competing second-order powers may deny any one state region-wide hegemony, as in the Middle East. With the exception of India and Brazil, second-order powers are unlikely to achieve hegemony over an entire geopolitical region. Rather, they can hope to exercise broad regional influence, with hegemony having practical significance only in relation to proximate states.

Third-order states influence regional events in special ways. They compete with neighboring regional powers on ideological and political grounds or in having a specialized resource base, but they lack the population, military, and general economic capacities of second-order rivals. Saudi Arabia, Libya, Taiwan, North Korea, Malaysia, Zimbabwe, Côte d'Ivoire, and Hungary hold such status. Lesser-order states such as Sudan and Ecuador have impact only on their nearest neighbors, whereas fifth-order states such as Nepal have only marginal external involvements.

Membership in the various orders is fluid. China is now a first-order power. It has gained economic strength through the opening of its system to world market forces, and its military strength has grown through expansion of its air power and its drive to create a "blue ocean" navy. A Russia restored to economic vitality and political cohesion that comes to a genuine political and military accommodation with China will not only reassert its major power status but may strengthen China's competitive position against Japan and the West as well. Accommodation between Russia and China is likely to be hastened by NATO's eastward expansion in the face of Russia's objections.

A decade ago, twenty-seven nations could be measured as potential second-order powers (Cohen 1982). Of these Saudi Arabia, Morocco, the newly renamed Democratic Republic of Congo, and Cuba have fallen from the ranking or never really attained it. The German Democratic Republic and Yugoslavia have disappeared from the map altogether. On the other hand, South Korea and Thailand have now achieved the rank of regional power. Among the most prominent regional powers that are now extending their influence to neighboring areas are Nigeria, South Africa, and Brazil. Third-order status is also ephemeral. Tunisia, Tanzania, Ghana, and Costa Rica have enjoyed and then lost such ranking with the waning of their ideological influence.

The impacts of major powers and second- and third-order states give regionalism increasingly important geopolitical substance. Ideologically "asymmetrical" states play special regional roles. They promote turbulence by challenging the norms of hegemonic regional structures and injecting unwelcome energy into the system. Sometimes this produces dialectic response that brings change in those regional norms—revolutionary Cuba, Titoist Yugoslavia, and market-oriented Côte d'Ivoire of the 1970s are examples (Cohen 1984).

The "Gateway" Concept: State and Regions

One of the most significant geopolitical phenomena of the past half century has been the proliferation of national states within the international system. In 1997, the number of sovereign states in the world was 193, of which 185 were members of the UN. Between 1993 and 1997, twenty new sovereign states emerged. This represents more than a quadrupling since 1939. Most of these states have emerged from colonialism as subsistence-based territorial units or, most recently, as products of the collapse of communist empires. State proliferation will continue because there are still forty-three political units in some form of dependency (colonies, dependencies, external territories) or trust and self-governing territories, protectorates, departments, or commonwealths. Mainly, these are small islands with restricted economic bases and physical isolation whose independence would not increase world conflict (see Table 3.1).

Furthermore, a considerable number of separatist or irredentist groups remain committed to military or terrorist struggles for freedom. There are more than thirty such candidates for independence (some, like Kosovo, are not likely to opt for independence but would join a neighboring state) (see Figure 3.4). For many, conflict will remain and contribute to short-term regional instability. Within and outside these two groupings, there are novel areas that are candidates for independence. These are Gateway states, uniquely suited to promote world peace and the more general rise of what has been called the "Trading State" (Rosecrance 1986).

The characteristics of Gateway states vary in detail but not in their overall context. Politically and culturally they are distinct historic culture hearths, with separate languages, religions, and educational standards, but with a common trait of having favorable access by sea or land to external areas (see Figure 3.5).

Economically, Gateways tend to be more highly developed than the core areas of their host state, for they are often endowed with strong entrepreneurial and trading traditions. When they are sources of out-migration, they acquire links to overseas groups that can provide capital flows and technological know-how. Small in area and population and frequently lying athwart key access routes, Gateways may be of military value to their host states, whose security needs require defense guarantees should the Gateways acquire political independence. The models for such states have existed historically in Sheba, Tyre, and Nabataea; in the Hanseatic League and Lombard city-states; in Venice; and in Trieste and Zanzibar. Andorra, Monaco, Bahrain, and Malta are modern-day versions. So were Lebanon and Cyprus before they were dismembered.

Located mainly along the border of the world's geostrategic realms and its geopolitical regions (and including islands that are or can be microstates or highly autonomous "statelets") or within an integrating Europe, Gateways states are optimally situated for specialized manufacturing, trade, tourism, and financial services, thus stimulating global economic, social, and political interaction. With independence, they help to convert zones of conflict to zones of accommodation—for example, Singapore and Hong Kong. The emergence of such states can facilitate the creation of boundaries of accommodation as foreseen by Lionel Lyde (1926) more than six decades ago.

The addition of substantial numbers of new Gateway states to the international system is in keeping with developmental theory because these are economically specialized states that help to link the system as a whole and its various parts. Much different from the traditional territorial unitary or federated states, whose goals included self-sufficiency and defense capaci-

TABLE 3.1 The World Geopolitical System in the Twenty-First Century

	Major Powers	2nd Order Powers (22)	Existing Gateway (17)	New Gateways (25)	New Post-Colonial (12)	New Rejectionists/ Separatists (31)	New Confederations (11)
North and Middle America	U.S.	Canada	–	Alaska, Bermuda, British Columbia, Hawaii, Quebec	–	–	–
		Mexico, Venezuela	Bahamas, Trinidad	Aruba/Netherland Antilles, North Mexico, Puerto Rico	Cayman Islands, French Guyana, Guadeloupe, Martinique, St. Martin	East Nicaragua	Colombia-Venezuela Westindia
South America	–	Argentina, Brazil				South Brazil	–
Maritime Europe & Maghreb	EU	Algeria	Andorra, Luxembourg, Malta, Mon, Finland	Azores, Catalonia, Northern Ireland, Midway Island, Vascongadas, Gibraltar	Canary Islands	Brittany, Crete, Greenland, Scotland, Sicily	–
Offshore Asia	Japan	Australia, Indonesia, South Korea, Thailand	Hong Kong, Taiwan, Singapore	Guam, Southwest Australia	American Samoa, French Polynesia, North Caledonia, North Mariana Island	East Timor, Ryukyu Island, Sulu Island/ Southwest Mindanao, West Irian	China-Taiwan
Heartland	Russia	Ukraine	–	Russian Far East	–	Chechnya, Crimea, Tuva, Yakutia	Slavonic Antata (Ru, Bel, Ukr, Kaz) Great Turkestan (Uz, Taj, Kir, Turk)

	China	Vietnam					
East Asia	China	Vietnam	–	Hong Kong/Shenzhen/Coastal Guangdong/Fujian	–	Tibet, Xinjiang	China–Taiwan
South Asia	–	India, Pakistan	–	Pashtunistan, Punjab, Tamil Eelam/Nadu	–	Baluch, Kashmir, Kerala, Nagaland, North Myanmar	Pakistan-Afghanistan
Middle East	–	Egypt, Israel, Iran, Iraq, Turkey	Bahrain, Cyprus	Mt. Lebanon, Palestine (West Bank/Gaza)	–	Khuzistan/South Iraq, Kurdistan	South Arabia–Gulf Syria-Lebanon-Iraq
Central and East European Gateway	–	Poland	Estonia, Latvia Finland, Slovenia	–	–	Transylvania, Trans-Dniestria	Baltics Bosnia-Croat-Serb
Subsaharan Africa	–	Nigeria, South Africa	Djibouti	Eritrea, Zanzibar	Mayotte, Réunion	Cabinda, Cape Provinces, North Somalia, Shaba, South Sudan	

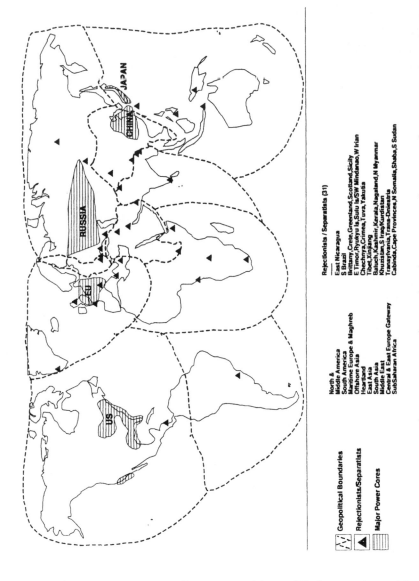

FIGURE 3.4 National sovereignty and geopolitical regions in the twenty-first century: Rejectionists and separatists

Geopolitical Boundaries

Rejectionists/Separatists

Major Power Cores

North &
Middle America
South America
Maritime Europe & Maghreb
Offshore Asia
Heartland
East Asia
South Asia
Middle East
Central & East Europe Gateway
SubSaharan Africa

Rejectionists / Separatists (31)

East Nicaragua
S Brazil
Brittany,Crete,Greenland,Scotland,Sicily
E Timor,Ryukyu Is.,Sulu Is/SW Mindanao,W Irian
Chechnya,Crimea,Tuva,Yakutia
Tibet,Xinjiang
Baluch,Kashmir,Kerala,Nagaland,N Myanmar
Khuzistan,S Iraq/Kurdistan
Transylvania,Trans-Dniestria
Cabinda,Cape Province,N Somalia,Shaba,S Sudan

ties, such states are mini-trading states with qualified sovereignty. They represent no military threat and are economic assets to their neighbors.

The ideal general system has countless numbers of parts or hinges that can connect with each other without having to move through rigidly controlled and hierarchical pathways. Similarly, a more flexible international system in which states are linked globally, regionally, and sectorially allows governments to cope more easily with shocks, as blockage points are by-

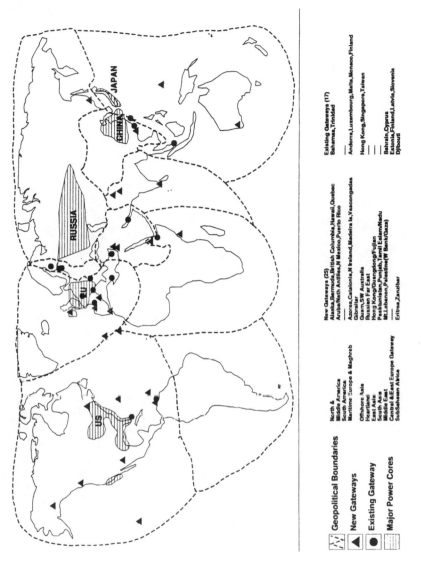

Geopolitical Boundaries

New Gateways

Existing Gateway

Major Power Cores

North &
Middle America
South America
Maritime Europe & Maghreb

Offshore Asia
Heartland
East Asia
South Asia
Middle East
Central & East Europe Gateway
Sub-Saharan Africa

New Gateways (25)
Alaska,Bermuda,British Columbia,Hawaii,Quebec
Aruba/Neth Antilles,N Mexico,Puerto Rico

Azores,Catalonia,N Ireland,Madeira Is,Vascongadas
Gibraltar
Guam,SW Australia
Russian Far East
Hong Kong/Guangdong/Fujian
Pashtunistan,Punjab,Tamil Eelam/Nadu
Mt Lebanon,Palestine(W Bank/Gaza)

Eritrea,Zanzibar

Existing Gateways (17)
Bahamas,Trinidad

Andorra,Luxembourg,Malta,Monaco,Finland

Hong Kong,Singapore,Taiwan

Bahrain,Cyprus
Estonia,Finland,Latvia,Slovenia
Djibouti

FIGURE 3.5 Gateways and geopolitical regions in the twenty-first century

passed and the system feeds on a multiplicity of nodes. Just as computer chips are made faster by making their circuitry more compact, Gateway states can make the world system more responsive by serving as its "mini-hubs." They can reach out to various first- and second-order powers and reinforce the unique character of each geopolitical region and also help bridge them to other regions. Uniquely suited to furthering peace, such novel states or autonomous economic entities such as Hong Kong can help fashion what Peter Taylor (1992) has referred to as a people-centered world map as a complement to the state-centered map.

Regions, as Well as States, Can Be Gateways

A promising potential geopolitical mechanism for restoring the balance between the Continental and Maritime realms is a Central and Eastern Europe Gateway region (see Figure 3.1). Such a region could facilitate the transfer of new energies into the faltering Russian core. Extending on the west from the Oder-Neisse Rivers and the Harz and Bohemian Mountains to the Northern Adriatic Sea, and on the east to the borders of the former USSR, a Central and Eastern European Gateway could be fully open to economic forces from the East and West. While the region's political and economic structures are adopting the West European models, its states will have to find a military posture that does not challenge Russian security goals.

Although demilitarization is not a viable option for the region, a form of "Finlandization" for much of Eastern Europe is a healthier option than either NATO or a resurrected Russian-dominated bloc. The addition to NATO of the Czech Republic, Hungary, and Poland is now a foregone conclusion. However, further expansion into Eastern Europe, such as into Slovakia, the Baltic states, Romania, Bulgaria, the Balkans or Ukraine, would be a strategic blunder of enormous consequence.

Another potential Gateway region is Middle America's Caribbean and Central American sector—the American Mediterranean (see Figure 3.2). Such a region would link South America, under Brazil's leadership, with the United States and Maritime Europe. From the standpoint of security, it would continue to lie within the orbit of the United States, a condition never realistically in doubt even when the USSR had footholds in Cuba and Nicaragua. However, the United States, with less reason to focus on military issues, can commit more of its resources to regional development, as can the other two geopolitical regions. Although Mexico directly represents a major focus for "offshore" American manufacturing, as well as a source for large-scale immigration, the American Mediterranean has the potential for attracting European, Japanese, and South American capital as points of entry to the American market, as well as U.S. capital. Puerto Rico can become an important hinge within the system should it become independent.

Conclusion

The world is rapidly progressing toward specialization and hierarchical integration, as its two geostrategic realms sort out the relationships of their respective internal power centers. Neither Russia nor China has yet to achieve the national self-confidence and mutual support that will enable the Heartland and East Asia to build a new chapter from the ashes of their schism, but they are beginning to do so. They are opening their national systems economically and, increasingly, politically. Meanwhile, the United States, Maritime Europe, and Japan have yet to agree upon a reallocation of global economic and military responsibilities.

Each of the world system's geopolitical regions is at a different stage of development, and thus their power and influence cannot be easily compared. The roles that second-order states play depend on the particular qualities that they bring to bear on their spatial and political-economic interaction with major powers and neighboring states. What helps to link the system is that the drive of the less mature parts to lower their entropic levels to ones already achieved by the more mature sectors is spurred by the latter's fear of the destabilizing effects of regional inequalities.

Sustained economic development means greater strength and self-confidence for the individual parts of a world system that since World War II has been characterized by attempts at top-down regulation. As the system progresses its parts become more open, more capable of drawing in new energies, and more likely to find balance through self-regulation—cooperation often being spurred by failure to achieve goals through war and competition.

In this chapter, I have set forth a global geopolitical system that is bounded by linking and balancing mechanisms—lowered entropy levels for some regions and raised levels for others, the rise of Gateways to strengthen the global network, and the attainment of a new strategic balance in Western Eurasia based on offsetting economic and military inputs between the Maritime and European Continental realms. The geopolitical system reflects not a New World Order—because that would imply a static, regulated, and precarious condition subject to violent swings and upheavals. Instead it suggests a new stage of dynamic equilibrium that is maintained in an ever-changing system through higher degrees of self-regulation and self-fulfillment of the parts. If the aftermath of the Cold War does not mean a world free from tension and chaos, it does mean a world in which the conflict is likely to be localized and of short duration. Perhaps it will lead to a new era of a "geography of accommodation" to replace the "geography of war" paradigm.

References

Cohen, Saul B. *Geography and Politics in a World Divided,* 2nd ed. New York: Oxford University Press, 1973.

_____. "A New Map of Geopolitical Equilibrium: A Development Approach." *Political Geography Quarterly* 1, no. 3 (1982): 223–242.

_____. "Asymmetrical States and Global Geopolitical Equilibrium." *SAIS Review* 4, no. 2 (1984): 193–212.

_____. "The World Geopolitical System, in Retrospect and Prospect." *Journal of Geography* 1 (January-February 1990): 1–12.

_____. "Global Geopolitical Change in the Post–Cold War Era." *Annals of the Association of American Geographers* 81, no. 4 (1991): 551–580.

Fairgrieve, James. *Geography and World Power.* London: University of London Press, 1915.

Hartshorne, Richard. "The United States and the 'Shatter Zone' in Europe." In *Compass of the World,* edited by H. Weigert and V. Stefannson, pp. 203–214. New York: Macmillan, 1944.

Kelly, Philip. "Escalation of Regional Conflict: Testing the Shatterbelt Concept." *Political Geography Quarterly* 5, no. 2 (1986): 161–180.

Lyde, Lionel. *The Continent of Europe.* London: Macmillan, 1926.

Mackinder, Halford J. "The Geographical Pivot of History." *Geographical Journal* 23 (1904): 421–444.

_____. *Democratic Ideals and Reality.* New York: Henry Holt, 1919.

Mahan, Alfred Thayer. *The Influence of Sea Power upon History: 1660–1783.* Boston: Little, Brown, 1900.

O'Loughlin, John, and Herman Van Der Wusten. "Political Geography of Pan-regions." *Geographical Review* 80 (1990): 1–19.

Rosecrance, Richard. *Rise of the Trading State.* New York: Basic Books, 1986.

Taylor, Peter. "Tribulations of Transition." *Professional Geographer* 44, no. 1 (1992): 10–12.

Von Bertalanaffy, Ludwig. *Organismic Psychology and Systems Theory.* Barre, Mass.: Clark University Press, 1966.

Werner, Heinz. *Comparative Psychology of Mental Development,* rev. ed. New York: International University Press, 1948.

Whittlesey, Derwent. *German Strategy of World Conquest.* New York: Farrar and Rinehart, 1942.

Chapter Four

International Boundaries: Lines in the Sand (and the Sea)

BRADFORD L. THOMAS

Plus ça change, plus c'est la même chose.
—Old adage

The world political map might well be likened to a giant mosaic depicting the world's geopolitical structure, with each of the individual pieces representing a separate State and the lines of mortar dividing them representing international boundaries.[1] The recent breakups of the Soviet Union, Yugoslavia, and Czechoslovakia seem to have introduced a phase of unprecedented geopolitical change—a shattering of some of the pieces and a rececementing of the shards with new lines of mortar, as it were.

Yet the geopolitical structure of the world has gone through other major upheavals and not so long ago. Between 1946 and 1986, ninety-five new States joined the world community, emerging from former dependencies in Africa, Asia, Oceania, and the Caribbean. In 1960 alone, seventeen new African States were added. Some of these were broken out of once larger and continuous pieces of the world mosaic, such as French West Africa; others were separate dependencies that passed intact into independent status. Among the recent breakups in Europe, that of the Soviet Union, in fact, can be viewed as simply a more recent phase in the postcolonial independence process.

A Dynamic Mosaic

Notwithstanding surges in the number of States as empires have broken up, the geopolitical mosaic in postcolonial African and Asian areas, once formed, has been fairly stable. The Latin American mosaic (with rare exceptions) for the most part has been stable since the beginning of the twentieth century. Geopolitical instability, in turn, has not been limited to former colonial areas. The mosaic of Europe, though stable from 1946 to 1990, had earlier gone through some wrenching changes—in States after 1919 and in their boundaries after 1945 (see Figure 4.1). These alterations continued a trend of periodic territorial adjustments between European States or empires throughout the nineteenth century as well as throughout history. Major adjustments were also made in the territory of the United States and its neighbors during the nineteenth century.

International boundaries were regarded by European scholars during the nineteenth and early twentieth centuries as fluid and changing phenomena, a concept clearly stemming from and supported by the incessant ebb and flow of the territorial expansion and contraction of States during that period. This aspect of international boundaries clearly lives on in the numerous disputes over boundaries that persist to this day. German political geographer Friedrich Ratzel regarded the State as a dynamic organism, of which the border zone was the epidermis that expanded as the State grew.[2] By contrast, however, the new nineteenth-century States of Latin America generally respected the colonial boundaries in place upon independence, although some conflicts have arisen nonetheless.[3]

After World War II, a trend developed in Europe, Africa, and Asia toward the preservation of existing international boundaries. The United Nations Charter in 1945 made territorial acquisition through warfare, which had contributed so much to instability of international boundaries, illegal under international law.[4] States newly emergent from former colonial status were the chief source of new international boundaries for several decades of the postwar period. Yet, even in these cases, there was a distinct preference for boundary stability, similar to the policies of Latin American countries a century and a half earlier. In 1964, the member States of the Organization of African Unity (OAU), in spite of having inherited the arbitrarily drawn boundaries of colonial powers that split ethnic and tribal groups into different States, issued a landmark resolution at Cairo pledging respect for the borders existing on their achievement of independence. This action was taken in recognition of the utter chaos that would ensue in attempting to alter Africa's boundaries. In both Africa and Asia, newly independent States have often based their claim to national territory on the administrative decrees of a colonial government. Even Iraq's claim to Kuwait in the recent Gulf War was based on its status under the Ottoman Empire.

FIGURE 4.1 *Central Europe at four stages through the twentieth century, show-ing the dynamic nature of the geopolitical mosaic over time.*

In 1975, in the Helsinki Final Act, the Conference on Security and Cooperation in Europe (CSCE, now the OSCE) called for the inviolability of international boundaries. And with some notable and unfortunate exceptions (such as the war between Armenia and Azerbaijan), this principle has governed national policies throughout Europe and the area of the former Soviet Union. It tempered a great many of the reemergent ethnic and territorial disputes that accompanied the disintegration of communism in Eastern Europe and, subsequently, the dissolution of the Soviet Union. A notable example is Hungary's official denial of a claim on the Transylvania region of Romania, where many ethnic Hungarians live.

The principle of inviolability of existing international boundaries has been extended to former internal administrative boundaries that have become international through the breakup of these States. Thus, republic boundaries in the former Soviet Union, the former Yugoslavia, and the former Czechoslovakia, and the provincial boundary between Eritrea and the rest of Ethiopia before 1993, became international boundaries when those internal administrative units attained independence.

Despite the general respect for existing boundaries in recent times, there have been some violent departures from the principle. In Africa, despite the OAU Cairo resolution, boundaries were violated when Ethiopia and Somalia fought a war in 1978 over Ethiopia's portion of the Ogaden region, when Libya occupied part of the Aozou Strip in Chad, and when Morocco occupied Western Sahara in 1975. The 1980–1988 war between Iraq and Iran and Iraq's 1990 attempt to absorb Kuwait are examples from Asia. In the Transcaucasus region, Armenia and Azerbaijan fought from 1992 to 1994 over the status of the Armenian enclave of Nagorno-Karabakh inside Azerbaijan (see Figure 4.2). After the 1992 breakup of Yugoslavia, ethnic Serbs and Croatians in Bosnia and Herzegovina and Serbs in Croatia attempted to carve off autonomous States.

The principle of self-determination of peoples that brought former colonies to independence, also enshrined in the U.N. Charter,[5] seems to be emerging as the chief potential source of any new international boundaries. The Russian Federation, itself the principal remnant of the breakup of the former Soviet Union, is experiencing civil unrest in some of the republics that it comprises. After a bitter civil war, the Republic of Chechnya has garnered increased, but not complete, independence from Moscow. The Republic of Tatarstan continues to exert pressure for greater independence. In Italy, northern Italians are pushing for an independent "Padania" separate from Rome, while a smaller group has come out for re-creation of the Venetian Republic dissolved by Napoleon. Europe is not alone. Farther east, ethnic-based separatism threatens China's western region and northern India. And in North America, the question of an independent Quebec still persists.

FIGURE 4.2   *Azerbaijan, showing the ethnically Armenian region of Nagorno-Karabakh, formerly an autonomous oblast of Azerbaijan and the object of recent armed conflict between Armenia and Azerbaijan.*

The disassembling of existing States contrasts sharply with preparations being made by the fifteen States of the European Union toward a common currency and the admittance of other States in Europe to the Union. The Union has already reduced or eliminated many traditional functions of its boundaries, such as customs and immigration checks for EU citizens, and these steps toward stronger economic and political integration could lead to an eventual political union of European States. Whatever the future holds, the relative lack of change in the world's geopolitical mosaic after World War II created an illusion of stability that has made the disintegration of familiar States all the more exceptional and unsettling. In a longer view of the history of international boundaries, however, stability would seem to be more the exception than the rule.

Lines in the Sand

All the breaking and remaking of the world's geopolitical mosaic has taken place under the basic principle of a single bounding line of cement between the pieces—the sanctity of sovereignty as defined territorially by international boundaries. The international community assumes that a finite line marks the limit of territorial jurisdiction of each State that is formed. Respect for that line is a condition of full membership in that community.

The idea of circumscribing the territory of a State with a finite line developed with the growth of nationalism and the concurrent evolution of the sovereign State as a political organism. These concepts, in turn, developed principally in seventeenth-century Europe and were exported around the world with the expansion of colonial empires. They were not originally part of most cultures over which the Europeans gained hegemony, but as non-European areas gained their independence from European colonial powers, they adopted the sovereign State concept. Boundaries are essential to the administration of the State power and territorial control that is vested in a national government.

Asian States in the past, like those in pre–seventeenth-century Europe, were based more on control over people, manifested as allegiance to a sovereign, than over territory. Although the center of the State could be easily identified, the periphery was amorphous and ill defined. In many cases, the sovereign's power radiated outward from the people over which he held absolute control to reach peoples or States that recognized his suzerainty—a looser and more distant form of control that often consisted principally of payment of tribute by the subjects. Some of the peripheral peoples or States recognized and paid tribute to two suzerains. The Asian "State" or, more properly, empire was thus more a phenomenon of gradation than of finite limits.

The loosely defined traditional nature of the Asian State also characterized the Ottoman Empire. The legacy of this ambiguity was evident recently in Iraq's claiming of Kuwait because Kuwait had been under Ottoman suzerainty between the 1870s and 1918 and was part of the Ottoman province *(wilayet)* of Basrah, which subsequently became part of Iraq. In 1899, however, Kuwait became a protectorate of Great Britain, an arrangement that ultimately led to its independence as a separate country in 1961. Kuwait's history testifies to the different levels of political control and allegiance that were traditional in the Arab cultural milieu.

A loose definition of territory does not work well in a modern world that requires administration of populations and resources. The lack of finite boundaries, often a legacy of nomadic herding societies, has been an impediment to the development of frontier area natural resources, such as oil and gas, that are often vital to raising the economic levels of the countries involved. In the Arabian Peninsula, finds of oil and gas resources have occurred in vast desert regions where undefined boundaries have been common. Several countries have now agreed on their boundaries and demarcated them on the ground. In the oil-rich Persian Gulf, a dispute between

Bahrain and Qatar over the Hawar Islands is before the International Court of Justice in The Hague, and Kuwait and Saudi Arabia are settling their offshore boundary through a producing oil field. A dispute persists, though, between Saudi Arabia and Yemen, who have not resolved a mostly undefined boundary through a large area of oil and gas exploration despite six years of talks (see Figure 4.3).

Concepts

The term *boundary* signifies the finite and often precise *line* surrounding and defining the territory of a State. In popular parlance, the words *border* and *frontier* are often used as synonyms, but these two terms have other, distinctive meanings. A border can mean both a boundary, as just defined, and the general zone within which the boundary lies—an *areal* concept. Sometimes, the phrase *border dispute* refers not to a disagreement over the location of the boundary but to a conflict over border functions between governments or over transgressions by groups of people on either side of an undisputed boundary line. And the term *borderlands* is often used to distinguish more clearly the areal concept of a border.

A *frontier* is also an areal concept, usually referring to a more loosely defined and perhaps wider zone than that indicated by the word *border*. It usually signifies the territory leading up to a boundary from both sides, sometimes representing a zone of gradation from one State to another in either jurisdiction or cultural attributes such as language. The idea of a frontier may be translated into legal applications of diminished jurisdiction, as with the *Zona Fronteriza* of Mexico, south of the border with the United States, where Mexican citizens resident in the zone have exemptions up to a certain value on the duties on certain types of goods purchased in the United States. Frontier zones may be set up as geopolitical entities of greater local autonomy for ethnic groups that extend beyond the boundary of the national territory, as with the autonomous North-West Frontier Province in present-day Pakistan that the British established along the border with Afghanistan, in an area occupied by the Pashtun people, who also constitute a major ethnic group within Afghanistan.

A frontier can be an area between two States where a precise boundary has never been defined, such as most of the area between Saudi Arabia and Yemen. In such regions, where precise boundaries have not been part of traditional nomadic tribal cultures, disputes have arisen in modern times over ownership of oil and gas resources. A frontier can also be the amorphous region that forms the margin of the settled and developed territory of a State, as in the American frontier of the nineteenth century. This notion of a frontier often carried with it the concept of a steadily expanding edge of national development and control. Although such frontiers may still be

76

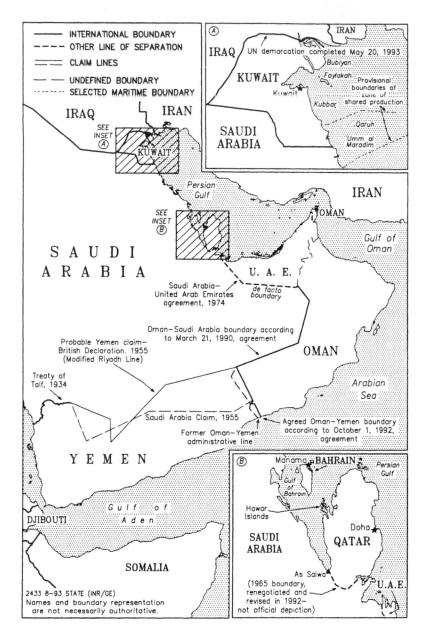

FIGURE 4.3   The Arabian Peninsula, showing international boundaries in several stages of evolution. Note the lack of an agreed boundary and the conflicting claims between Saudi Arabia and Yemen over most of their frontier. Determination of this boundary has been the subject of talks between the two States over the past six years, without resolution.

found in some land areas (e.g., the Amazon Basin), the newest frontiers of national expansion are in the sea, where many coastal States have set out zones of maritime jurisdiction within which they exploit resources in and under the sea.

An international boundary is usually established through a treaty or agreement between the two States that it separates. The treaty may simply provide an *allocation* of territory in principle between the two States, but it more often contains a *delimitation,* that is, a description of the boundary alignment in a text or by marking on a map. It may also provide for a boundary commission to execute a *demarcation,* which is the precise location of the alignment in the field and the marking of it with pillars or monuments. The commission is customarily given authority to make minor adjustments in location for realities on the ground. Not every boundary that is agreed upon and honored between States is demarcated; many boundaries, for reasons of history or geography, remain delimited but not demarcated (for example, the boundary between Iraq and Kuwait, prior to 1993).

A boundary agreement does not come into force until it has been ratified by each government and the instruments of ratification have been exchanged. Ratification may follow the agreement quickly or take several years. States may elect to honor the agreed boundary in the interim, even though the agreement is technically not in force. Where a final boundary has not yet been agreed upon, temporary or provisional lines are often established pending the negotiation of a final treaty, the ratification of an agreement, or the results of an arbitration or adjudication. Other lines of national separation like this are often referred to as *de facto,* as opposed to *de jure* (or legal) boundaries.

Boundary Functions

International boundaries have several *functions.* Not only do they mark the territorial extent of a State's power, but they also serve to control the movement of people and goods—or even ideas—from one State to another. In this age of widespread travel and shipment by air, similar "boundary" functions have to be exercised at international airports. The controls over the movement of people have several objectives, such as restricting entry of criminal or terrorist elements, people with highly infectious diseases, economic migrants whose presence might depress part of the labor market, or in some cases in the past, people whose racial or cultural origins are not totally acceptable to the national population. They have also served, in the case of totalitarian States, to prevent the exit of people seeking freedom or better economic opportunities elsewhere. Often special arrangements are instituted to permit workers to enter a State for a limited time. In frontier areas, special permits are sometimes issued to workers on one side of a boundary to commute across it daily to employment on the other side.

Restrictions on the movement of goods across boundaries are more strictly economic in motive. Often they are designed to protect domestic industries from the "unfair" competition of goods produced more cheaply in other States. The resulting tariffs and customs duties imposed at borders may also provide a convenient source of hard currency income. They can also serve to keep out harmful products such as illegal narcotics, illegal arms, or infected animals and plant matter. In addition, border controls are needed, in some cases, to prevent the unwanted export of especially valued items such as hard currency or gold bullion, or of sensitive technology that might be used to a State's detriment by a potential enemy. Control of ideas can be achieved partially by confiscation of "objectionable" printed matter at border check points, but it is much more difficult to achieve, despite radio or television jamming activities, in this day of global electronic communications.

American geographer S. Whittemore Boggs, writing in 1940, felt that the functions of boundaries were "in general negative rather than positive. To at least some degree they restrict the movement of peoples and the exchange of goods, of money, even of ideas. . . . International boundaries are intended to serve protective functions of various kinds. They can not promote trade or human intercourse as is sometimes contended."[6] Today, with the development of a global economy and the disappearance of Cold War tension, many of the world's States, through bilateral trade agreements or multilateral federations, are gradually eliminating or reducing the number of boundary functions. Most of these efforts are aimed at removing tariff or other economic barriers to free trade, as with the North American Free Trade Agreement (NAFTA) or South America's MERCOSUR. The most prominent and advanced example, perhaps, is the gradual, painstaking movement by members of the European Union toward not only economic but social and political integration, eliminating controls on movement of people and labor across boundaries as well as restrictions on movement of goods and capital. At the same time, pressures from refugee flows and illegal migrants have led to stiffened immigration controls, even among States otherwise reducing boundary functions, and continued illegal narcotics smuggling prompts more intensive efforts to seal borders against such movements.

International boundary functions vary greatly around the world as barriers to movement. The reader of a world map on which usually there is one uniform symbol for agreed and ratified international boundaries must be aware that, although the lines symbolized may be of equal political status, they may be quite dissimilar in regard to functions.

Disputes

Bounding of States with finite lines often leads to disputes over the location of the boundary. Disagreements of this kind between States can range from purely technical differences over the precise alignment of the boundary, sometimes even within an agreed delimitation, properly called a *boundary dispute*, to claims over pieces of territory, large or small, properly called a *territorial dispute*. There is no clear-cut definition of the point at which one becomes the other, that is, how much ground is involved to warrant calling a dispute "territorial" rather than "boundary." Although a territorial dispute inherently means disagreement over the boundary, the difference is more one of concept and origin rather than size of area involved.

Boundary disputes originate more from differences of interpretation of treaty terms or of the ground features identified in a boundary delimitation, for example, where does the line actually lie in a wide river chosen as a boundary, or which mountain ridge in a cordillera was intended as the boundary by the treaty framers. Many such disputes have arisen because of ambiguous or vague identification in early treaties of terrain features chosen as the location of the boundary. This type of dispute has often resulted from inadequate exploration or mapping of the frontier area at the time of negotiation. In many cases, negotiators failed to understand the need not only for accurate mapping prior to delimitation but also for precise geographic references in drafting the boundary treaty. The vague pre-1993 boundary delimitation between Iraq and Kuwait is a classic example.

Territorial disputes, on the other hand, more often originate in political differences such as historical claims on lost lands or irredentist policies promoting union of ethnic groups separated by a boundary. An example of a boundary dispute would be that between Qatar and Saudi Arabia, whereas Iraq's claim on the entire State of Kuwait would be an example of a territorial dispute (see Figure 4.3). Disputes over the sovereignty of islands are also examples of territorial disputes.

Sometimes disputes have been both boundary and territorial. For example, Ecuador has traditionally laid claim to vast areas of the Amazon Basin east of the Andes Mountains stretching all the way to the Amazon River, an area held by Peru. Although Ecuador has continued recently to make vague claims about "sovereign access" to the Marañón and Amazon Rivers, current boundary discussions with Peru are focused on specific alignment disagreements (termed *impases*) in several sectors of a boundary that was delimited in the 1942 Protocol of Rio de Janeiro that ended a 1941 war between them (see Figure 4.4).[7]

A similar situation has occurred between Russia and China. Although China has historically made territorial claims to vast areas of Siberia al-

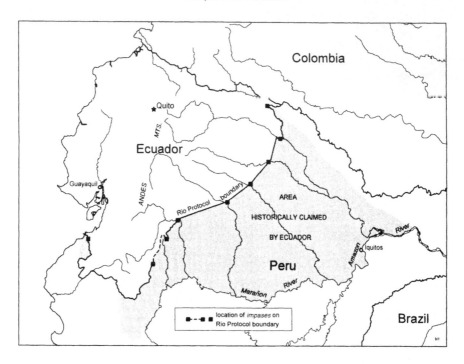

FIGURE 4.4   *Ecuador, showing (1) the extent of its historic claim as depicted on its official maps and (2) the 1942 Rio Protocol boundary with Peru and the locations along it of the specific impases that are the subject of current bilateral talks. The map does not show Ecuador's claim, which it included on one of its impases, to "sovereign access to the Marañón-Amazon."*

legedly lost in earlier wars with Russia, it signed a 1991 treaty with the then Soviet Union settling the eastern sector of their border in the rivers claimed by the Soviet Union as the boundary. The Soviet government accepted the middle of the main channel of those rivers as the boundary alignment and negotiations focused solely on identifying that main channel among several intertwining channels forming the rivers.

The dispute between Iraq and Kuwait has vacillated between boundary and territorial. At times Iraq has disputed the existence of Kuwait as an independent State, claiming it as part of Iraq, as it did during its 1990 invasion of the country. At other times, apparently recognizing Kuwait as a separate State, Iraq has disputed the location of their boundary—making repeated attempts to gain control over two islands that would allow it to control both sides of an important waterway to one of its main ports.

Boundary and territorial disputes range in intensity from those that generate actual warfare to those handled through diplomatic channels. Sometimes, two States that have peaceful and friendly relations in every other sector of their affairs will still be engaged in a bitter boundary or territorial dispute. Only those disputes officially put forward through diplomatic channels by legitimate governments of independent States qualify as true disputes. Any others, whether voiced by unofficial dissident groups, individuals with political agendas, academic writers, or journalists are not officially disputes. Moreover, not every official disagreement over a boundary may be properly termed a "dispute." A boundary location may be questioned in principle, for example, but the issue not actively pursued. There are many places in the world where such disagreements have lain latent for years because neither State cares to disturb otherwise neighborly relations with the other. In still other cases exact boundary locations remain uncertain simply because the two States have never gotten around to delimitation or demarcation. Disputes over such things as border functions or over the actions of groups of people along a border, rather than the alignment of the boundary itself, are not *boundary* disputes.

When two States disagree over boundary or territorial issues, there are several mechanisms that can be used to resolve them. The simplest, and the preferable method in most cases, is bilateral negotiation. This procedure requires first the political will of both governments to settle the issue in this manner. If this will exists, the parties can settle their dispute in any way that is mutually satisfactory. If coming to the negotiating table is particularly difficult for either or both parties, they may call in a third party to facilitate the negotiations by providing good offices, conciliation, or mediation. If these various approaches to bilateral negotiation are not feasible, there are more formal, legal third-party options that produce binding resolution. These require the two States to agree to submit the dispute and to abide by the decision that results. One method is to submit the dispute to a "neutral" party for arbitration. Argentina and Chile have used this method on several occasions, submitting their boundary disagreements either to the King or Queen of England or, in a later case, to the pope for a decision. Another method is to submit the dispute to an international arbitration tribunal usually composed of judges or legal experts from the two States involved plus others from "neutral" States. Egypt and Israel, for example, settled a 1980s boundary disagreement following Israel's withdrawal from the Sinai Peninsula by submitting it to a tribunal comprising representatives from Sweden, Switzerland, and the United States, in addition to those from Egypt and Israel.

A third option is to submit the dispute to the International Court of Justice in The Hague, Netherlands, for adjudication. El Salvador and Honduras, as part of a 1980 treaty ending warfare between them, used this method to settle a lingering dispute over various pockets of disputed territory along their border. Libya and Tunisia used the court to settle a dispute over their maritime boundary in the Mediterranean Sea, and Canada and the United States also used the court to settle a dispute over their maritime boundary in the Gulf of Maine. Indonesia and Malaysia are planning to submit a dispute over islands in the Celebes Sea to this court.

With disputes essentially territorial in nature, resolution, especially with arbitration or adjudication, usually involves application of one of the five recognized principles in international law by which a State can acquire sovereignty over territory: occupation, prescription, conquest and annexation, cession, and accretion or avulsion. *Occupation* refers to effective control over a piece of territory by a State. If the area was not previously under the sovereignty of any State, it is not enough, as in earlier centuries, to proclaim discovery. The discovery must be followed by some symbolic act of incorporation and by actual occupation and use of the territory. *Prescription* occurs when a State exercises peaceful, unopposed, continuous governance over a territory actually belonging to another State. Effective opposition to this situation by the legally sovereign State does not require military action; it is enough to issue diplomatic protests to the State presuming to exercise rights of prescription.

Conquest and annexation had been viewed as valid up to the end of World War II. The changes in borders and territory made as result of that war, mostly by the Soviet Union in Europe (see Figure 4.1) were generally accepted by the world community,[8] but, as noted earlier, the United Nations Charter in 1945 made this method of territorial acquisition invalid and the 1975 Helsinki Final Act essentially reaffirmed this principle in its call for the inviolability of international boundaries. It was violation of this U.N. charter provision that brought about the coalition of member nations united against Iraq when it invaded Kuwait.

Cession occurs when there is a transfer of territory from one country to another, whether voluntary or involuntary. It is usually accomplished by a treaty. The 1867 purchase of Alaska from Russia by the United States is an example of peaceful cession. *Accretion and avulsion* allow for the acquisition of land by the action of natural forces that cause such things as changes in the courses of boundary rivers or expansion of deltas into the sea. Accretion is the gradual addition of land through such processes; avulsion is a more radical addition through a sudden action, as in a flood that causes a major change in a river's course or a volcanic eruption that adds to an area's territory with flows of new material. More recently, boundary treaties have been written to take these factors into account.[9]

Lines in the Sea

The idea of a single, finite line bounding national territory has been considerably diluted in recent times with coastal States. As these States increasingly turn their attention to capturing the resources in and under the sea, maritime boundaries have been drawn to mark off which parts of the sea or seabed pertain to which coastal State. In earlier times, the shoreline of a coastal country and a narrow band of controllable water just off it marked the edge of national jurisdiction. Now, however, maritime limits established under the 1982 United Nations Convention on the Law of the Sea (LOS Convention) have created a set of zones of progressively diminishing national jurisdiction moving seaward from the shore.

The right of States to claim maritime zones well precedes the 1982 LOS Convention. The concept of a *territorial sea* dates back to the seventeenth century. The *continental shelf* entered the realm of international law with the proclamation by President Truman of the U.S. continental shelf in 1945. In 1958, the Geneva Conference on the Law of the Sea produced four conventions, two of which—on the territorial sea and contiguous zone and on the continental shelf—codified for the first time the nature and extent of maritime zones. The next couple of decades saw a surge of claims to maritime zones by coastal States. The Third U.N. LOS Conference during the 1970s prompted an all-time high in maritime claims, many now including the more comprehensive *exclusive economic zone* (EEZ) concept developed by the conference, even before the 1982 LOS Convention that resulted was signed. By 1990, there were eighty EEZ claims.[10] With the coming into force of the LOS Convention in 1994, there was another flurry of activity, either of maritime claims or of further ratifications.

Sea Frontier Concepts

The term *maritime boundary* refers, in general, to the *limits* of the maritime zones established off the shores of coastal States for administration and resource exploitation. The rules governing these zones were codified not only in the 1982 LOS Convention but also in earlier conventions in 1958. In addition to separating one zone from another, these limits separate the maritime jurisdictions of neighboring States, whether adjacent to one another along the same coast or facing one another across a body of water. Except for the *territorial sea,* they are not, in the strict sense, *international boundaries.* Only the territorial sea limit marks the edge of a State's full sovereign powers, subject only to provisions of *innocent passage* of all foreign vessels. In the other zones, the State exercises only limited economic sovereignty.

The *baseline* from which the territorial sea and the other limited jurisdiction zones are measured is normally the mean low water mark of mainland or island shorelines. The maximum seaward extent of the *territorial sea* allowed by the LOS Convention is 12 nautical miles from that baseline.[11] Although a State's territorial sea may not legally exceed that width, lesser claims are permissible.

A State may employ *straight baselines* along the coast in certain geographical situations specified in the LOS Convention: across deep indentations in the coastline or between fringing islands off the coast. Inland from such straight baselines is a regime of *internal waters* in which the sovereignty of the State is no different than on dry land, except for allowance of innocent passage if such waters have not been enclosed before (see Figure 4.5). Archipelagic States such as Indonesia or the Philippines are entitled to enclose the waters within their islands as *archipelagic waters* by establishing a set of *archipelagic straight baselines* connecting the outer edges of the outermost islands.[12]

Seaward from the territorial sea, a State may claim a *contiguous zone* not to exceed 12 nautical miles, the purpose of which is to allow the State to prevent or punish infringements of its customs, fiscal, or sanitary laws pertaining to the territorial sea. Beyond the territorial sea and its contiguous zone are several varying kinds of maritime jurisdiction over resource exploration and exploitation. The broadest is the exclusive economic zone in which a State controls resources both on or under the seabed and in the water column above it, their conservation and management, protection and preservation of the marine environment, scientific research, and the erection and use of artificial structures. Other States cannot carry on resource exploration or exploitation or any of the other activities reserved to the coastal State without that State's express permission. The outer limit established by the LOS Convention for the EEZ is 200 nautical miles.

In other resource exploitation zones beyond the territorial sea, the jurisdiction of the coastal State is more restricted, limited to fishing in the *fishing zone* and to resources on or under the seabed (oil or gas under the seabed but also bottom-dwelling sea life on the seabed) in the *continental shelf*. Although the fishing zone is not specifically covered by the LOS Convention and it may be of any breadth, customary practice limits it to 200 nautical miles. The outer edge of the continental shelf is where the submerged prolongation of the continental landmass drops down to the deep ocean floor. This drop-off may occur beyond 200 nautical miles, in which case a coastal State can claim a continental shelf up to 350 nautical miles if the submarine topography so indicates. If a State's physical continental shelf does not extend as far as 200 nautical miles, the LOS Convention allows the State to claim that breadth for continental shelf jurisdiction.[13]

On the high seas, in EEZs or fishing zones, or over continental shelves, all vessels and aircraft enjoy customary high seas rights of navigation (including submerged) or overflight, and of laying and maintenance of submarine cables and pipelines, and other activities associated with them. In territorial seas, foreign vessels are permitted a more restrictive *innocent passage,* which, in general, means continuous and expeditious navigation, not stopping or anchoring except as incidental to ordinary navigation, nec-

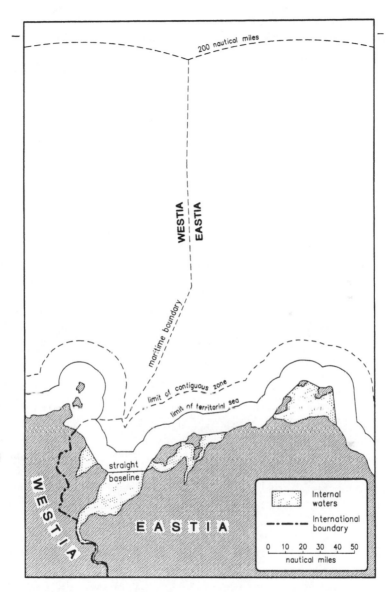

FIGURE 4.5   *Diagram illustrating the concept of maritime limits. The 200-nautical-mile line could be the outer limit of a continental shelf, a fishing zone, or an exclusive economic zone (EEZ), depending on the law of the coastal State. Where fewer than 400 nautical miles separate coastal States facing each other across an enclosed sea, a mutually agreeable maritime boundary must be negotiated. Where straight baselines are used to connect outer islands of an insular State, they are called archipelagic straight baselines, and the internal waters created by them are called archipelagic waters.*

essary because of distress, or needed for rescue operations, and not submerged or launching aircraft. However, in international straits connecting one area of high seas or EEZ with another but so narrow that the territorial seas on either side merge, the LOS Convention allows for *transit passage*. This kind of passage permits all rights of high seas navigation and overflight. *Archipelagic sea lanes passage,* in which all ships and aircraft are also allowed high seas navigation and overflight, applies to sea lanes that are generally used for international traffic in what otherwise would be archipelagic (i.e., internal) waters enclosed by archipelagic straight baselines. Innocent passage is still allowed in other parts of archipelagic waters.

Seaward Expansion of National Jurisdiction

The expansion of coastal States' jurisdiction into the oceanic realm in the post–World War II era is giving new dimensions to a traditional land-based definition of national territory. The LOS Convention, which came into force after ratification from the required sixty countries was achieved in November 1994, has set the nature and limits of this offshore jurisdiction, but only the territorial sea represents an offshore zone of full State sovereignty (reduced only by provisions of innocent passage for foreign vessels, as noted). Over four-fifths of the world's coastal States (including the United States) now claim the maximum 12-nautical-mile territorial sea.[14] Gradually, countries that have previously claimed territorial seas of greater breadths are adjusting those claims to the 12-nautical-mile standard and many (such as the U.S.) whose previous territorial sea claims were smaller have extended them out to 12 nautical miles. In the latter cases, there has been a seaward extension of national territory, in the full sovereignty sense of that concept.

Beyond the territorial sea, the convention provides extended maritime zones of "economic sovereignty," such as the EEZ or continental shelf, but in these zones State jurisdiction is limited to control over exploration and exploitation of marine resources. Although a popular perception appears to be growing that these extended zones represent an extension of national territory out to 200 nautical miles (or even beyond, with some continental shelves), the waters in them are, in fact, international waters, with attendant high seas rights of navigation. On the other hand, of course, exclusive national control over marine resources *has* been gained out to EEZ or continental shelf limits.

No changes are envisioned to the provisions of the LOS Convention for maritime zones at this time, so the offshore territorial sovereignty of coastal States extends to 12 nautical miles, with a further extension of more limited, strictly economic jurisdiction outward to 200 nautical miles. In the case of some continental shelves that extend, under the convention's definition, beyond 200 nautical miles, economic jurisdiction over resources on or under the seafloor reaches farther. Around the world (with the exception of Antarctica) there are twenty-nine such places that have been identified.[15]

From a purely resource perspective, in fact, there is no longer any part of the sea floor that is not under either national or international jurisdiction. Under the LOS Convention, exploitation of the mineral resources of "the Area," the remaining deep ocean floor beyond the 200-nautical-mile limit or the twenty-nine extended continental shelves of greater breadth, is to be administered for the common good of all the world's States by the International Sea-Bed Authority. This body, consisting of an Assembly, an Executive Council, and a Secretariat, will also have an "Enterprise" that, once the economic viability of seabed mining is demonstrated, will actually engage in seabed mining on behalf of developing States unable to finance such costly operations.

Island Sovereignty Disputes and the Law of the Sea Convention

Disputes over islands and rocks, many hitherto unknown, have recently surfaced and gained worldwide attention. Although these sovereignty disputes themselves are not new, bilateral relations have been strained by public posturing and an unwanted international publicity of what had been "back burner" issues. Conflicting island claims have been around for decades, sometimes even for centuries, and range the globe, with particular concentrations in East Asia and the Middle East (see Figure 4.6).

Many of these disputed islands have little or no inherent value. Some are no more than rocks jutting out of the sea, or minuscule coral cays or atolls with little surface vegetation and no fresh water. In earlier times, some of them actually did have some intrinsic value—as transitory bases for fishermen, as sites for navigational aids, as refueling and provisioning stations, for strategic control of shipping lanes, or for the resources found on them, such as mineral deposits or bird droppings (guano). Today, it is not the islands but rather the potential resources in and under the seas around them that have given them renewed importance. The difference is the LOS Convention that gives these islands strategic value in providing for a territorial sea of up to 12 miles around any land feature that projects above high tide and allowing for EEZ or continental shelf jurisdiction out to 200 nautical miles around that feature as long as it is not a rock that "cannot sustain human habitation or economic life of [its] own."[16] The LOS Convention has served as a catalyst for States to review their offshore islands, leading to a resurgence of island disputes.

The LOS Convention by itself, however, has not brought about this situation. Geopolitical interest in maritime jurisdiction in general, as well as that with respect to islands, originates in the gradual dwindling or deple-

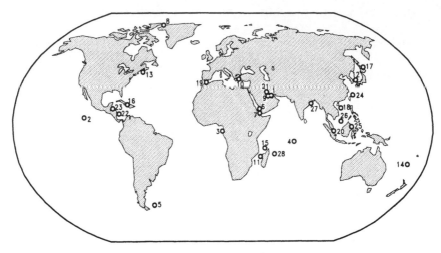

Dispute

1 Abu Musa and the Tunbs
2 Clipperton Island
3 Corisco Bay, several islets
4 Diego Garcia
5 Falkland Islands/Islas Malvinas, South Georgia and
  South Sandwich Islands
6 Farasan Islands (southern part)
7 Hanish Islands
8 Hans Island
9 Hawar Islands
10 Imia (Kardak) Rocks
11 Juan de Nova, Bassas da India, Europa, Glorioso and
  Tromelin Islands
12 Liancourt Rocks (Takeshima / Tok Do)
13 Machias Seal Island / North Rock
14 Matthew and Hunter Islands
15 Mayotte
16 Navassa Island
17 "Northern Territories"
18 Paracel Islands
19 Peñon de Alhucemas, Peñon de Velez de la Gomera,
  Isle Chafarines
20 Pulau Batu Puteh, Pulau Pisang
21 Qaruh and Umm al Maradim
22 San Andres y Providencia
23 Sapodilla Cays
24 Senkaku Islands (Diaoyu Tai)
25 Sipadan and Ligatan Islands
26 Spratly Islands
27 South Talpatty Island
28 Tromelin Island

Disputants

Iran - United Arab Emirates
France - Mexico
Equatorial Guinea - Gabon
Mauritius - United Kingdom
Argentina - United Kingdom

Saudi Arabia - Yemen
Eritrea - Yemen
Canada - Denmark
Bahrain - Qatar
Greece - Turkey
France - Madagascar

Japan - South Korea
Canada - United States
France - Vanuatu
Comoros - France
Haiti - United States
Japan - Russia
China - Vietnam
Morocco - Spain

Malaysia - Singapore
Kuwait - Saudi Arabia
Colombia - Nicaragua
Belize - Guatemala - Honduras
China - Japan
Indonesia - Malaysia
China - Malaysia - Philippines - Vietnam
Bangladesh - India
France - Seychelles - Mauritius - Madagascar

FIGURE 4.6   *Map of the world showing the locations of currently known island disputes. Each symbol represents either a disputed single island or a disputed group of islands.*

SOURCE: International Boundaries Research Unit, *Maritime Briefing* (University of Durham, U.K., 1998).

tion of land-based resources and the growing need of the world's States to develop offshore food and energy sources to meet the demands of burgeoning populations and economies. The development of the LOS Convention was a response to this rising interest in the oceans and represented an attempt to set rules and standards for the exploitation of those offshore resources. Moreover, technological improvements in exploiting maritime resources more efficiently and economically or at greater and greater depths into the oceans have reinforced the geopolitical interest.

Regardless of the maritime resource nature of most island disputes, the LOS Convention *does not* contain any provisions that discuss the resolution of disputes over any territory, be it islands or mainland. Although the LOS Convention provides for several bodies for adjudicating disputes and for a commission to oversee claims to continental shelves beyond 200 miles, it does not deal with sovereignty issues. The LOS Convention addresses only the establishment of maritime jurisdiction zones and is premised on the assumption that a particular State has undisputed title over the territory, including islands, from which the maritime zones are claimed.

Disputes over territorial sovereignty of islands must be settled before maritime zones can be delimited. At least four prominent island disputes appear to be heading for such arbitration or adjudication: in East Asia, Malaysia has agreed with both Singapore and Indonesia to submit their respective island disputes to the International Court of Justice (ICJ); Bahrain and Qatar may be settling their dispute over the Hawar Islands in the ICJ; and Eritrea and Yemen have agreed to take their dispute over the Hanish Islands to arbitration by an international panel.

If the interest of the disputing parties is less about the piece of territory itself and more on the potential maritime jurisdictional area to be generated by it, the island sovereignty dispute can be put on hold and some other means devised to utilize the maritime resources around the island. Such finessing of the sovereignty question by seeking a solution based on resource considerations is usually driven by commercial interests in the marine area—be it fisheries or, more likely, oil and gas development. Oil companies, for example, are reluctant to invest money in areas in which there is no clear title. If a workable scheme can be devised whereby resource development can occur and benefits flow to both countries, then the immediate sovereignty issue is minimized. Joint development areas, in which resolution of a maritime boundary is usually left for a future date, are one such solution, and several of these have been established successfully.

The Airspace Dimension

In addition to being a line on the earth's surface, an international boundary is also a two-dimensional plane that extends from that surface both to the center of the globe and to the top of the atmosphere (the exact limit of which, however, has not been defined). Projection of the boundary downward determines rights to subterranean mineral or hydrocarbon deposits in the frontier area. Projection upward creates airspace above each State that is also presumably inviolable. The world's States have developed a cooperative system for civilian airliners to use in serving international passenger and traffic routes. Serious problems do arise, however, when States refuse permission to enter or even transit that space. Unauthorized entry into a State's declared sovereign airspace had tragic consequences in 1983 when Soviet fighters shot down a Korean airliner that was alleged by Soviet authorities to have intruded into their airspace. And the U.S. bombing of Libya in retaliation for its sponsorship of terrorism was complicated by airspace transit denials that required the U.S. planes to fly a more circuitous route through the Strait of Gibraltar.

Offshore, the outer limit of the territorial sea, being the edge of a State's sovereignty, is also the outermost limit of its national airspace,[17] with the exception of international straits covered by territorial seas where the LOS Convention, as noted earlier, allows for *transit passage*. This kind of passage permits all rights of high seas overflight (hence the use of the Strait of Gibraltar by the U.S. planes headed for Libya). A similar Convention provision, *archipelagic sea lanes passage*, permits high seas overflight over sea lanes normally used for international navigation and overflight in archipelagic waters or in the territorial sea just outside the baselines.

Conclusion

The world's geopolitical mosaic continues its historic pattern of shifting and changing. The source of most of the traditional changes—the conquering of one State or part of a State by another—is phasing out, in part under the influence of the U.N. Charter and the provisions of the Helsinki Final Act. Yet States continue to split or merge, and new States, such as Palau, which achieved full independence from U.S.-administered U.N. trusteeship in October 1994, are still being created.

With the changing pattern of the geopolitical mosaic, the concept of a single bounding line between its pieces has subtly changed, especially among the coastal States intent upon capturing the resources of the marine environment lying off their shores. There, instead of a simple, sharp termination of a State's jurisdiction, we see an outward gradation of zones of diminishing jurisdiction. The brightly colored coastal pieces of the mosaic do not end abruptly but essentially fade away through a broader zone, which, on a map, could be shown with paler colors.

Continuing reduction in restrictive boundary functions will also affect all of us in the next century. To the extent that States shed some of their traditional border functions, we could see a blurring of the sharp, finite land boundaries similar to that evidenced in the gradations of different marine jurisdictional zones. Will international boundaries of the future have to be depicted on maps in different shades of hue to indicate their different levels of function? Moreover, the trend toward greater autonomy of certain geopolitical units within some sovereign States and the possible emergence of supranational governments (such as the EU) could create several levels of "sovereignty," as opposed to the current system of 190 independent States. How will the cartographers of the future depict the Russian Federation, for example, if certain local republics within it become more autonomous? How will they depict Italy, if it breaks down into several "independent" States tied in a loose federation? How will they depict the geopolitical structure of a new Europe united politically as well as economically? Designers of future world maps will have to become very clever in depicting a dynamically changing mosaic of spaces bounded by lines of quite variable significance.

Notes

1. See E. J. Soja, *The Political Organization of Space,* Commission on College Geography Research Paper no. 8 (Washington, D.C.: Association of American Geographers, 1971), p. 9. Throughout this chapter, *State* (capitalized) refers to a sovereign, independent, territorially defined political entity, as opposed to *state*, referring to a sub-State territorial unit. *State* is considered synonymous with *country, nation* (in the United Nations sense), *nation-State, sovereign State,* or *independent State.*

2. Friedrich Ratzel, "Die Gesetze des räumlichen Wachstums der Staaten," *Petermanns Mitteilungen,* vol. 42 (1896), p. 102.

3. The principle is known as *uti possidetis juris.*

4. *Charter of the United Nations . . . ,* Article 2, paragraph 4.

5. Ibid., Article 1, paragraph 2, and Chapter XI.

6. S. Whittemore Boggs, *International Boundaries: A Study of Boundary Functions and Problems* (New York: Columbia University Press), p. 11.

7. This boundary was subsequently demarcated by a binational commission over approximately 90 percent of its course.

8. One exception was the refusal of the United States to recognize the 1940 annexation of Estonia, Latvia, and Lithuania by the Soviet Union.

9. The Israel-Jordan Peace Treaty of 1994 has specific provisions for accretion or avulsion of the Jordan River, which forms part of the boundary.

10. Robert W. Smith, "Summary of Maritime Claims as of May 13, 1997," U.S. Department of State, Office of Ocean Affairs, unpublished factsheet.

11. A nautical mile is equivalent to 1.15 statute miles or 1,852 meters.

12. Articles 46 and 47 of the LOS Convention define an "archipelago" and detail standards for archipelagic straight baselines.

13. Article 76 of the LOS Convention gives a detailed definition of the outer margin of the continental shelf.

14. Robert W. Smith, "The State Practice of National Maritime Claims and the Law of the Sea," paper presented to the conference on "State Practice and the 1982 Law of the Sea Convention," Cascais, Portugal, 1990, Table 1, Graph 1.

15. Victor Prescott, "National Rights to Hydrocarbon Resources of the Continental Margin Beyond 200 Nautical Miles," paper presented to "Boundaries and Energy: Problems and Prospects," the 4th International Conference of the International Boundaries Research Unit, University of Durham, UK, 18–19 July, 1996.

16. Article 121. The LOS Convention does not precisely define capability of sustaining human habitation or economic life. In acceding to the LOS Convention in July 1997, the U.K. declared Rockall—a rocky pinnacle from which it had claimed a 200-mile fisheries zone—not capable of sustaining human habitation and therefore not entitled to an extended maritime zone and announced its intention to redraw U.K. fisheries limits. This appears to be the first official rejection of a specific feature as eligible for extended maritime zones under the Article 121 criteria.

17. Greece has been an exception to this principle in claiming an airspace of 10 nautical miles that exceeds its 6-nautical-mile territorial sea. Some seventeen countries with excessive territorial sea, hence airspace, claims, even out to 200 nautical miles in eleven cases, have yet to roll them back to the 12-nautical-mile U.N. standard (Robert W. Smith, "Summary of Maritime Claims as of May 13, 1997").

References

Allcock, John B., et al., eds. *Border and Territorial Disputes.* Harlow, U.K.: Longman, 1992.

Boggs, S. Whittemore. *International Boundaries: A Study of Boundary Functions and Problems.* New York: Columbia University Press, 1940.

Brownlie, Ian. *African Boundaries: A Legal and Diplomatic Encyclopaedia.* Berkeley and Los Angeles: University of California Press, 1979.

Charney, Jonathan, and Lewis Alexander. *International Maritime Boundaries.* Dordrecht, Netherlands: Martinus Nijhoff, 1993.

Degenhardt, Henry W. *Maritime Affairs—A World Handbook: A Reference Guide to Maritime Organizations, Conventions, and Disputes and to the International Politics of the Sea.* Harlow, U.K.: Longman, 1985.

Geopolitics and International Boundaries. A journal edited by the Geopolitics and International Boundaries Research Centre, School of Oriental and African Studies, University of London, and published by Frank Cass and Company, London.

Hartshorne, Richard. "Suggestions on the Terminology of Political Boundaries." *Annals of the Association of American Geographers* 26 (1936): 56–57.

International Boundaries Research Unit (IBRU), University of Durham, U.K., publishes *Boundary and Security Bulletin, Boundary and Territory Briefings,* and *Maritime Briefings.* See the IBRU website at [http://www-ibru.dur.ac.uk]. IBRU also maintains a very useful worldwide e-mail list for exchange of information on international boundaries called *Int-boundaries.* For information contact IBRU at [ibru@durham.ac.uk].

Jones, S. B. *Boundary Making: A Handbook for Statesmen, Treaty Editors, and Boundary Commissioners.* Washington, D.C.: Carnegie Endowment for International Peace, 1945.

_____. "Boundary Concepts in the Setting of Place and Time." *Annals of the Association of American Geographers* 49 (1959): 241–255.

Kristoff, Ladis K. D. "The Nature of Frontiers and Boundaries." *Annals of the Association of American Geographers* 49 (1959): 269–282.

Pearcy, G. Etzel. "Boundary Types." *Journal of Geography* 64, no. 7 (1965): 300–303.

_____. "Boundary Functions." *Journal of Geography* 64, no. 8 (1965): 346–349.

_____. "Dynamic Aspects of Boundaries." *Journal of Geography* 64, no. 9 (1965): 388–394.

_____. *World Sovereignty*. Fullerton, Calif.: Plycon Press, 1977.

Prescott, J.R.V. *Maritime Political Boundaries of the World*. London and New York: Methuen, 1985.

_____. *Political Frontiers and Boundaries*. London: Allen & Unwin, 1987.

United States, Department of State. *Limits in the Seas*. Washington, D.C., various dates, 1970–present. An ongoing series of studies of agreed bilateral maritime boundaries or unilateral maritime zones. Studies 106 and following available from the Office of Ocean Law and Policy, U.S. Department of State. Studies 1 through 105 available for consultation in the Geography and Map Division, Library of Congress, and other libraries.

United States, Department of State. *International Boundary Studies*. Washington, D.C., various dates 1961–1985. A series of studies of agreed bilateral international land boundaries. Available for consultation in the Geography and Map Division, Library of Congress, and other libraries.

Chapter Five

The Power and Politics of Maps

ALAN K. HENRIKSON

The "World" as a Map

For an appreciation of the complexities of international relations, at both the academic level and the professional level, an understanding of cartography is more important than is commonly realized. "The world we have to deal with politically," as Walter Lippmann wrote many decades ago, "is out of reach, out of sight, out of mind. It has to be explored, reported, and imagined." Like denizens of Plato's Cave, officials who are for the most part immured in their government bureaucracies must experience the events of the vast external realm indirectly. The "persistent difficulty" of statesmen, noted Lippmann, who had been an adviser at the conference in 1919 that produced the Versailles Treaty, "is to secure maps on which their own need, or someone else's need, has not sketched in the coast of Bohemia."[1]

Foreign policy, which comprises the spectrum from information to trade to defense and security policies, is based on statesmen's conceptions of the actual world and their own countries' situations within it. No less a political figure than Napoleon asserted that "the policy of a state lies in its geography."[2] In this political context, "geography" means more than topographical irregularity or climatic variation, more than distribution of mineral resources or configurations of human settlement, and even more than patterns of commerce and finance or networks of transport and communication. It signifies the almost a priori spatial frame of reference, which is usually centered on an imagined point of origin within the core area of a country from which the activities of that society are organized and proceed. The foreign thrusts as well as the domestic initiatives of a regime and nation are thus guided.

In order to have a political plan, statesmen must have a geographical conception, which requires the cartographic image of a map. Actual maps, as well as mappable ideas and spatial consciousness and sensitivity ("mental maps"), are a critical variable—occasionally even the decisive factor—in the making of public policy. Without knowledge of what is going on geographically and, to help organize that knowledge, an actual "picture" or structured image of the developments and events taking place at large, it is not possible to conceive of or, even less, to carry out effectively a large-scale purpose. No military campaign, development project, or even, at a more abstract level, diplomatic strategy or information program can be intellectually sustained or practically executed unless it is plotted spatially. The conception, development, and implementation of policy depend in complex and interrelated ways on mapping.

Through maps and charts, through these graphic representations virtually alone, the earth and its regions have acquired the shapes by which we know them. Not only the outlines but also the sizes, distances, and directions of the globe's territorial and maritime expanses are most often apprehended first through the logical structures and vivid images of cartography. Long before they are actually visited, the world's remote geographical areas usually are visualized, sometimes very fancifully, in the mind's eye.

The principal sources of our world vision, and much of our factual knowledge of the earth and the happenings upon it, are the terrestrial map, the nautical chart, and, increasingly, the high-altitude aerial photograph and satellite image. Indeed, our very concept of "world," which is an ideological construct that is usually more philosophical than geographical in content, can be framed and articulated by cartography.

Images such as the sublime "Earthrise" photograph taken from the Moon during the *Apollo 8* voyage at Christmas in 1968 have transformed our sense of place and belonging. Men and women, otherwise divided along lines of nationality, ideology, race, and class, could suddenly be seen as one humankind.[3] At more technical levels of observation, too, important insights were gained into Earth and its conditions. These were facilitated by new possibilities of space-based mapping, which used remote sensors and computers. With this advanced technology and the development of geographic information system (GIS) software, vast areas of the planet could be scanned at once and masses of collected data regarding resources, habitats, and communities could readily be translated into geometrically correct, flexible, detailed cartographic forms.[4]

Maps are not only formal projections that net the world mathematically. They are products of geographical thought that typically present themselves visually. Maps have, first of all, a *synoptic* quality. They display conditions that exist contemporaneously over the entire area covered, making it possible to see "everything at once"—affording a general view as well as an opportunity to examine specific features more closely. The "simultaneous complexity" of the events in the perceptual array of a map is perhaps cartography's dominant and most distinguishing attribute.[5] Maps also have what has been called a *hypnotic,* or suggestive, effect. "Cartohypnosis," as a former U.S. State Department geographer has termed the subtle persuasiveness of maps, causes people to "accept unconsciously and uncritically the ideas that are suggested to them by maps."[6] Finally, maps can be *emblematic,* or representative in a pictorially symbolic way of a territory and the polity or other entity or occurrence upon it. Older maps, particularly, tend to be festooned with the indicia of statecraft, navigational technique, or the fruits of discovery or conquest. Map emblems have appeared in historical paintings—portraits of Napoleon, for example—as territorial symbols and/or geographical icons. Maps thus may be embedded in the discourse of politics and of art, just as political symbols can be embedded in the language of maps.[7]

Map representations usually are made, though they need not necessarily be, on paper. Clay tablets were used in ancient times. Oxhide or vellum (fine-grained lambskin, kidskin, calfskin) was employed from then through the Middle Ages. The beautiful large *mappa mundi* (c. 1283 A.D.) in Hereford Cathedral, for example, is drawn on leather. The mapping of exploration in the Renaissance period—the great works, many of them collected in atlases, of Martin Waldseemüller, Gerardus Mercator, Abraham Ortelius, and Jodocus Hondius—were visually spectacular.[8]

Much earlier cartography was ephemeral, produced for temporary or even one-time use. Religious pilgrims on journeys to holy places in Europe or the Near East carried road maps on cloth or paper strips. North American Indians made maps on birch bark or merely drew them in the sand, indicating that they did not consider maps to be valuable in themselves but rather to be mere accessories to geographic knowledge. The heavens as well as earth can be mapped. The Pawnee apparently used star charts painted on and perforated through elkskin for orientation and probably for contemplation.[9] Indigenous American maps usually served immediate, practical needs such as locating sites, showing directions, or tracing routes. Unlike most European and European-American maps, which emphasized boundaries, Indian maps typically were composed of small circles, for villages, and interconnecting lines, for pathways. There were no vast rectangular frames to record property lines, land grants, or territorial claims.[10] Native American cartography, being neither legalistic in motivation nor aimed at future historical validation, probably was not even meant to last. The Indians' real maps, the timeless, cosmological ones, as well as the transitory, utilitarian ones, were mostly in their heads.

Such "mental maps," or noncartographic maps, should be, but rarely are, included within any consideration of mapping. "Cognitive" maps, as they sometimes are also called, arguably are "the ultimate maps" because they are the ones actually relied upon in making decisions regarding the environment and movement within it.[11] It can be misleading, however, to term such cognitive knowledge of environmental reality "maps" unless such cognizance involves conscious spatialization. That is to say, there must be held in the mind not merely a multitude of geographical facts but also a more or less accurate sense of the voluminousness of the world and of the places within it that are known in roughly correct relationship to each other. There should be not only cognitive territory, in other words, but also some isomorphism, or similarity in structure, between the mental map and the physical world.[12]

The structures of mental maps can be highly complex, but they have a number of basic common features. Adapting terms from the work of the planner Kevin Lynch, who applied them only on an urban scale, one may distinguish the following constituent elements of many mental maps. These maps consist of *paths,* or the channels along which a person regularly travels or imagines doing so; *edges,* or the internal and external boundaries that inhibit his movements; *districts,* or the geographical-cultural regions in which he resides or within which he might move; *nodes,* or the intersections on which his and others' activities center; and *landmarks,* or the geographical signs he refers to for self-orientation and perhaps also for the direction of others.[13] The geographical knowledge reflected in a mental map ought to be capable of being transposed, in its essential elements and pattern, from the mind to a surface—that is, it must be capable of being actually drawn or otherwise cartographically rendered.

What drawn or printed maps do, that brain- or computer-stored geographical data bases as such do not, is to represent clearly an *image* of the environment. Such images are useful, and sometimes even necessary, for the communication of spatial-geographical relationships and facts to others. They may even be helpful in self-orientation, although they need not always be consciously so called upon. Even in an inattentive state, skillful behavior in space still may be possible under the guidance of well-articulated schemata.[14]

Mapmaking, including mental mapmaking, and map using, which includes map perceiving and map contemplating, are both visual and cognitive. Vision and thought—visual thinking—constitute a unified mental process.[15] It is through the lens of the map, cartographic and noncartographic, that we see, know, and even create the larger world.

Map Geometry, Map Symbols, and Their Perception

The notion of "map," in both broad and narrow senses, is essentially that of a model, a representation of a geographical area (usually) on a flat sur-

face. It ordinarily has the further characteristic that each point on the cartographic diagram corresponds to an actual geographical position on the
earth, according to a definite scale or system of projection. Map projections
transform the curved, three-dimensional surface of the earth onto a two-dimensional plane. Flattening inevitably produces distortions—consider what
happens to an orange peel when flattened on a table top. Hence, as an old
saying warns, "All maps lie flat, therefore all flat maps lie."

In general, scale distortion increases with distance from a standard point
or line. The mapmaker minimizes distortion by centering the projection in or
near the region to be featured by the map. The usual trade-off that has to be
made in mapmaking is that between what is called conformality and equivalence—that is, between preserving the shapes of the earth's configurations
and keeping areas everywhere equal so that there is no size variation anywhere on the map. Except for a globe, which is often made up of stretched
paper map-gores, no map can be perfectly conformal *and* equivalent.

The developable, or flattenable, surfaces most commonly used for map
projections are the plane, the cone, and the cylinder (see Figure 5.1). Projection of the earth grid of parallels (latitude) and meridians (longitude) onto a
tangent plane gives a gnomonic projection, which is especially useful in
long-range aerial and other navigation. On this azimuthal map, Earth's
shapes become highly distorted as the distance along the radiating azimuths
from the center of projection (point of tangency) increases. All great-circle,
or shortest-distance, routes, however, can be represented conveniently as
straight lines.

Projections onto a cone produce maps on which many elements—shapes,
distances, angles—are true, but this is so only along the line of tangency, or
standard parallel. Away from that line, distortion develops. Conic maps
have been used for showing areas of vast east-west extent, such as the territory of the former Union of Soviet Socialist Republics.

Projections upon a cylinder, often favored for world maps such as Gerardus Mercator's famous conformal projection, are not faithful to area. High
above and far below the line of tangency at the equator, the sizes of lands
and seas become greatly exaggerated. The classic illustration of the Mercator map's faults is the island of Greenland, which is actually about the same
size as Saudi Arabia but is, in appearance, larger than South America (see
Figure 5.2). The wonderful utility of the Mercator projection lies in the
realm of maritime navigation because, since a line between any two points
on it gives true direction, a sailor's compass course can easily be drawn as a
straight line. If such a compass route (rhumb-line or loxodrome) lies near
the equator or nearly on a meridian, the difference between it and the
shortest-distance, great-circle route is comparatively small. As most shipping occurs in the middle latitudes, the advantages and benefits of the Mercator projection thus are considerable.[16]

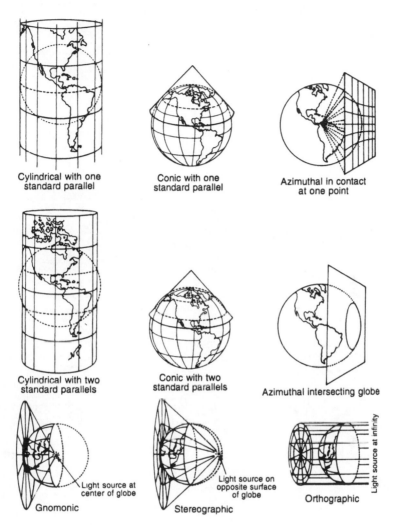

FIGURE 5.1   *Geometrical properties of various cylindrical, conic, and azimuthal projections.*

SOURCE: Norman J. W. Thrower, *Maps and Man: An Examination of Cartography in Relation to Culture and Civilization* (Englewood Cliffs, N.J.: Prentice-Hall, 1972). Reprinted by permission.

For the purpose of showing, in a meaningful comparative way, the distribution of elements upon the earth, such as mineral resources, agricultural products, or population, it is necessary to use equal-area maps. An equal-area map that presents the earth without rounding or interrupting is the so-called Peters projection (see Figure 5.2), advocated during recent decades

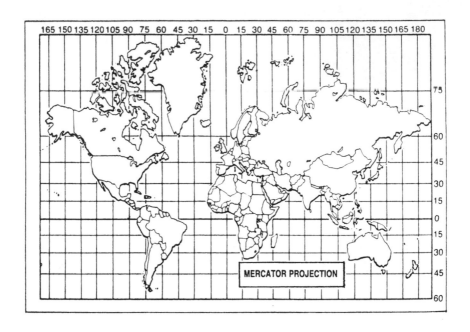

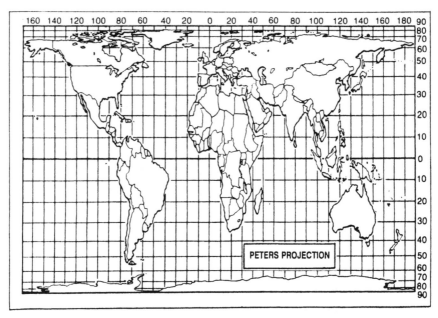

FIGURE 5.2   *Conformal and equal-area world maps.*

FIGURE 5.3   The Asia-oriented medieval worldview.

by the German historian Arno Peters.[17] Such is the distortion of continental outlines and political boundaries on this rectangularly framed map, however, that Arthur Robinson, a senior academic cartographer interested in the "look" as well as the technical qualities of maps, has remarked of the Peters map that "the landmasses are somewhat reminiscent of wet, ragged, long, winter underwear hung out to dry on the Arctic circle."[18]

It is not only the principle of geometrical projection that influences the perception of maps. The orientation of the map—that is, the direction the earth is turned on a map—also is a powerful determinant of the way the world is imagined and studied. Seeing "North" traditionally at the top of global maps can lead an uncritical map viewer to assume, without realizing the subtle effects of this convention, that northern territories, peoples, and even cultures are higher—superior qualitatively as well as cartographically. As is well known, certain societies—early Egypt, for example—did not invariably place North at the top. For Egyptians, perhaps even today, the course of the Nile River suggested a downward, South-to-North cartographic flow. In medieval Christian maps, which were highly schematic, the East, or Asian part where "Eden" or paradise was fancied to be located, was on top, at the head. The form of these maps was circular—an O partitioned by a T, representing the world's three principal waterways—the Tanais (Don) and Nile Rivers across, and the Mediterranean downward (see Figure 5.3). Jerusalem, the "navel" in this figurative world body, was usually placed at the center of these T-in-O designs. The large thirteenth-century map in Hereford Cathedral was a culmination of this tradition.[19]

The North-on-top convention, for world maps especially, probably originated with the Greek scientist-geographer Claudius Ptolemy. North was placed at the top in Ptolemy's maps because the better-known localities of the world—and, as we now know with more certainty than he did, the majority of the earth's actual land areas—were in the northern latitudes. World maps with North at the top were easier to study.[20] Nonetheless, the North-on-top image of the world should not be considered unalterable, as the producers of a popular series of "Turnabout" maps, with Australia or South America at the head, cartographically point out.[21]

In order to overcome the rigid North/South or East/West cardinal-direction thinking, the noted cartographer Richard Edes Harrison, trained as an illustrator and architect, urged viewers of his maps to accustom themselves to seeing the earth's patterns from *any* direction. Because of the hypnotic effect of the Mercator projection and the convention of North being situated at the top of the page, he believed, Americans and others had compromised a critical capacity of visualizing the world in different ways. He sought to make the earth's forms, in any position on a map, as recognizable and as familiar as a french curve on a drawing board.[22] Not only Harrison's standard-projection maps but also his wonderful bird's-eye-view, or perspective, maps of the various theaters of World War II demonstrated brilliantly the virtues of cartographic agility.[23]

The centering of maps is another cartographer's variable, which can be used to demonstrate the primary importance of the place that is at the focus. Indeed, one of the unfortunate consequences of colonialism and the condition it engendered, still affecting attitudes in some parts of the Third World today, is a feeling that "the center" lies elsewhere, usually in the metropolitan capital of the erstwhile imperial power. The objective realities on which such a feeling may be grounded, and the sensation of peripheralness itself, cannot be altered, of course, simply by shifting or reducing the graphic frame of the map. Nonetheless, it is noteworthy that one of the first steps of a newly independent country often is to commission a national atlas, to print stamps with a map of the country's outline on them, and otherwise to use the emblem of the map to assert the country's new identity in a new setting—a new pride of place.[24]

The centering of maps can also, alternatively, be used to demonstrate "encirclement," suggesting vulnerability, weakness, and uncertainty. Before and during World War I, Imperial Germany proclaimed its fear of *Einkreisung,* that is, of being ringed about by hostile powers intent upon choking off its access, limiting its opportunities, and blighting its future. The German Imperial Fleet, for example, was said to be "bottled up" by the British Royal Navy. A generation later, propagandists in Nazi Germany complained of the same discriminated-against geographic condition. By force of arms, especially naval but also aerial and land units, the German war machine sought to break through the enclosing circle of the Allies' power. Some of the Third Reich's maps illustrated this *Durchbruch,* or break-out, attempt.[25]

The U.S.-led Western policy of "containment" of the Soviet Union and its presumed effect on the Russians is roughly analogous, even in the encircled party's tendentious use of maps. During the U.S.-Soviet summit meeting off Malta in December 1989, for example, the Soviet leader, Mikhail Gorbachev, handed President George Bush a blue and white map allegedly showing the Soviet Union's "encirclement" by U.S. bases as well as American aircraft carriers and battleships. Prepared by Marshal Sergei Akhromeyev, this map was designed to underscore the Soviet General Staff's contention that U.S. vessels armed with submarine-launched cruise missiles posed an especially serious threat to the U.S.S.R. For a moment, according to a detailed account of this episode, President Bush was at a loss for words. President Gorbachev then said tartly, "I notice that you seem to have no response." Bush answered by pointing out to Gorbachev that the Soviet landmass was shown on the map as a giant white empty space, with no indication of the vast military complex that U.S. forces were intended to deter. "Maybe you'd like me to fill in the blanks on this," he said. "I'll get the CIA to do a map of how things look to us. Then we'll compare and see whose is more accurate."[26]

Besides geometry, which, as noted, includes issues not only of projection but also of perspective, coverage, and centering, it is symbolization that determines the nature and the effectiveness of a map. Apart from the basic question of the selection of map features—that is, the thematic content of a map, the characteristics of the earth that are to be observed and described—"symbolization" refers to the kind of graphic indicators used. In a comprehensive listing of "map symbols" produced by the cartographer Mark Monmonier, they are representative markings that may vary in size, shape, graytone value, texture or surface pattern, orientation or direction, and hue or color attribute. Shape, texture, and hue are effective in showing qualitative differences, as among land uses or dominant religions, he notes. For quantitative differences, size is more suited to showing variation in amount or count whereas graytone is better for portraying differences in rate or intensity. The symbols that vary in orientation are most useful for "representing winds, migration streams, troop movements, and other directional occurrences."[27]

The most obviously "dynamic" of map symbols are the oriented, or directional, ones such as the concentric circles and arrows drawn on the German maps to indicate the pressures of *Einkreisung* constraint and the counterforces of a *Durchbruch* strategy. Even the contrasting of colors and the juxtaposition of filled and voided areas, as in the Soviet map designed to support the argument of Western "encirclement" of continental Russia, stimulate interest. The psychological effects of variations of color and other symbols on maps have been much studied.[28]

It is not only propaganda maps, many of which are patently purposive and "readable," that can have an energizing effect on a viewer. Nearly all maps, because of the systemic variations and unique elements in them, are dynamic to some degree. As the cognitive psychologist and art theorist Rudolf Arnheim maintains, the shapes and colors of cartography have an "animating" quality. What meets the eyes first and foremost when looking at a map, he explains, are not the measurable phenomena represented by dimensions, distances, or the geometry of shapes on the map but "the expressive qualities" carried by the stimulus data. Studying a map can actually make the viewer feel the underlying spatial forces of the map structure "as pushes and pulls in his own nervous system."[29] Indeed, it is the "look" of maps that is decisive. If a map does not engage the mind of the active viewer and enter into a reciprocal exchange of the mapmaker's ideas and information and the viewer's understanding and interests, it does not succeed in achieving its purpose as a map. It is, in that case, not a map but rather an inert collection of lines, dots, and blank spaces, perhaps mixed with color.

The processes of map perception never will be completely analyzable by purely scientific measurement of the map/viewer interchange. The geometrical and symbolic elements of maps must combine to communicate concepts and images, and these are and must be perceived as wholes, as mentally integrated *sets* of sensations and induced feelings. The whole-forming qualities, or *Gestaltqualitäten,* of maps are grasped by the unifying intuition rather than simply through an accumulation of stimulus-response reactions.[30]

Power, Politics, and Policy

Maps are purposeful and persuasive, sometimes explicitly and nearly always implicitly. "Every map is someone's way of getting you to look at the world his or her way," proposes Lucy Fellowes, co-curator of the Cooper-Hewitt National Museum of Design exhibition "The Power of Maps." Part of the reason why maps are so effective, further suggests Denis Wood, the other co-curator of the exhibit, is "the impression they manage to convey that they are precisely above such interest. They are convincing because the interest they serve is masked."[31]

The patronage of maps repays thought.[32] Cartography, as the sociologically inclined historian of mapping Brian Harley pointed out, was always "the science of princes." In the Muslim world, the caliphs and then the sultans were known to have sponsored mapmaking, as did the Mughal emperors of India. The rulers of China also did so. In early modern Europe, absolute monarchs lent their patronage to cartography. From the eighteenth century, when national topographic surveys began, state sponsorship increasingly took over, favoring the policies and interests of elites.[33]

The map is almost the perfect representation of the state. It evenly covers the territory of a country.[34] It hierarchically organizes it, with the capital—Paris, London, Berlin, or Washington, D.C.—symbolically at the center, and provinces, counties, Länder, or states on the periphery marked down through the use of symbols as inferior orders of government.[35] Beneath these, there are still-lower orders of civic organization. The existence of some social and cultural realities—ethnic clusters or religious centers, for example—might not be recognized on state-sponsored maps at all.

Some of the "silences" on maps, as Harley calls the blank spaces on them, are "silences of uniformity," of standardization. Others are more specifically intended, the results of deliberate exclusion, willful ignorance, or even actual repression. The removal or the alteration of place-names—the named locations of conquered peoples or minority groups, for example—creates eloquent "toponymic silences."[36] The state projection of geometrical designs throughout a country in the form of straight-line jurisdictional boundaries, transnational highways, and preserves of one kind or another—thus establishing "order upon the land"—can also produce what might be termed geographical silences, or the structural subordination of the natural landforms that shape human communities. Cartography, like politics itself, was and remains today, in Harley's characterization, "a teleological discourse, reifying power, reinforcing the *status quo,* and freezing social interaction within charted lines."[37]

The teleological thrust, or natural purposefulness, of state-centered cartography extends also into the external realm. Exploration into unknown parts, and the cartographic recording thereof, was sponsored by Portugal's Prince Henry the Navigator and also by subsequent monarchs or presidents of other countries, including Spain, France, Great Britain, the Netherlands, and the United States. American expansion, no less than that of European empires, was preceded by mapmaking and geographical imagining. President George Washington himself was a surveyor. His fellow Virginian Thomas Jefferson became a continentwide and global plotter. His Lewis and Clark Expedition (1803–1806) measured the width of the continent and opened it to the course of "manifest destiny," aimed at establishing an empire on the Pacific.[38]

The attainment by the United States of a broad position on the Pacific slope, in consequence of its military victory over Mexico in the 1846–1848 war, caused it for the first time to be placed at the center of world maps.[39] Other factors influencing this Americentric cartographic adjustment were the Lieutenant Charles Wilkes's U.S. Exploring Expedition into the Pacific (1838–1842), the discovery of gold in California (1848), the restoration of

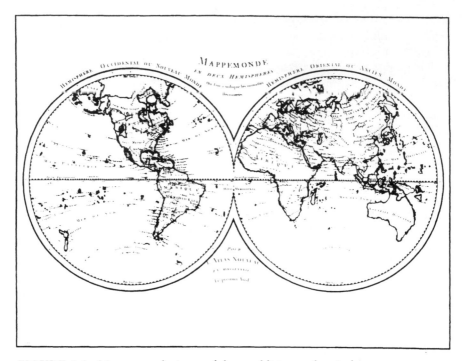

FIGURE 5.4   Mappemonde *(map of the world)* in two hemispheres.

the Union with the North's military success in the Civil War (1865), and the American purchase of Alaska from Russia (1867). The U.S. victory in the Spanish-American War of 1898 confirmed the centrality of the United States in the world order. On American-produced world maps, such as the "Map of the World on the Mercator Projection" shown in S. Augustus Mitchell's *New General Atlas* and subsequent maps, the whole global frame shifted—the world moved![40]

In U.S. foreign policy at that time, it was not, in truth, the world as a whole but, rather, the "Western Hemisphere"—a cartographer's convention before it became a geopolitical concept—that seemed appropriate for American concerns. The 1823 Monroe Doctrine had postulated that the European and American hemispheres were "essentially different" systems. The future of the New World ought therefore, it seemed, to proceed independently, in clear political separation from the Old World.[41] The cartographic template for the "two spheres" in international politics was the European split-hemisphere *mappemonde* dating from the Renaissance era (see Figure 5.4).[42]

Since the nineteenth century, until at least the time of World War II, the notion of "hemisphere defense" was the touchstone of American national-

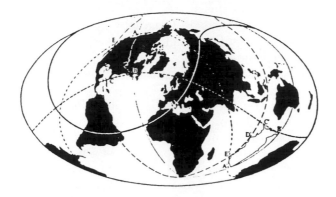

FIGURE 5.5   *The sum of all hemispheres containing all of the United States.*

security thinking. This was in part a subtle consequence of a prevailing mental picture of the United States being located in a separate hemisphere or at the center of the world shielded by two vast oceans. This cartographic logic either removed the "Eastern Hemisphere" from any contact with the New World altogether or split the mighty Eurasian landmass in two. The Europeans' Far East ended up on the far west of American world maps. Efforts to "dehypnotize" the American popular thinking by demonstrating graphically, through a series of alternatively centered maps, that the "Western Hemisphere" idea was arbitrary, and that "hemispheres" including the United States could be defined that would cover all of Europe and indeed most of the rest of the globe, probably in themselves did not have much effect (see Figure 5.5).[43] It took the Japanese attack on Pearl Harbor from across the Pacific and the conversion of the Atlantic Ocean by U.S. Navy forces into an "American lake," joining America and Europe, to disabuse the country of the notion of continental or hemispheric impregnability and safety.[44]

Despite the post–World War II alliances between the United States and Western Europe as well as Japan, the concept of "this hemisphere" as a relevant policy idea continues to exert an intellectual and emotional force in America. President George Bush's "Enterprise for the Americas" initiative of June 1990, which was based on President Ronald Reagan's earlier "North American Accord" proposal of November 1979 favoring Canadian-U.S.-Mexican cooperation, is a recent case in point. The role of geographical imagery in engendering a sense of North American or even Pan-American diplomatic coordination and comity can be crucial. Whether the successfully negotiated North American Free Trade Agreement (NAFTA) succeeds in fostering a sense of North American "community" may depend partly on whether the Canadian, American, and Mexican peoples have a

strong enough, and sufficiently shared, mental map of North America as a
unified entity.[45] The use of actual cartography to define and describe that
landmass, which for three centuries after Columbus had no political dividing lines and whose natural geological-economic "grain" runs mainly north
and south, could enhance NAFTA's chances of larger success, beyond a mutual sharing of markets.[46] The same proposition—the indispensability of a
vibrant geographical concept and image of their common region—may also
govern the outcome of future economic discussions between North American and South American countries.[47]

From the experience of World War II, an entirely different, and cartographically novel, image of the future international relations of the United
States developed. This was the notion of an "Arctic Mediterranean" space,
which synoptically gathered all the lands, seas, and also air spaces around
the Arctic Circle and outward from it in every direction in a new world order of peace, friendship, and commerce, particularly by means of aviation.
The base map for this alternative image, often touted as a replacement for
the equator-based Mercator projection, was the North Pole–centered global
chart, such as the widely reproduced azimuthal equidistant world map prepared by Richard Edes Harrison (see Figure 5.6).

Harrison's map, rotated 90° so as to situate Europe (the prime meridian
running through Greenwich, England) rather than the continent of North
America upon the vertical axis, or spine, of the world body, was used in de-
signing the map emblem of the new United Nations Organization. This UN
symbol of pacific universalism, which expressed Americans' postwar hopes
if not their realistic expectations, was officially described as follows: "A
map of the world representing an azimuthal equidistant projection centered
on the North Pole, inscribed in a wreath consisting of crossed convention-
alized branches of the olive tree; in gold on a field of smoke blue with all
water areas in white."[48] No political boundaries are indicated on it. Today,
with the military-strategic barriers between East and West having been
overcome, that idealistic map image can perhaps become practically relevant.[49]

Polar projections increasingly have been used to demonstrate the oneness
of the world in more scientific ways as well. Satellite imagery, often produced from observations by spacecraft that repeatedly circle the earth in
polar orbits, is so synoptic as to be capable of revealing global, systemic
patterns. The visual emergence of the "ozone hole" over Antarctica in the
late 1980s is a case in point. The actual picture of the "hole," shown in the
striking false-colors of the widely publicized satellite imagery of the phenomenon, contributed powerfully to international recognition of the problem and to public support for a negotiated solution.[50] So, too, did graphically presented information regarding ozone depletion over the more
heavily populated higher latitudes of the Northern Hemisphere. The Sep-

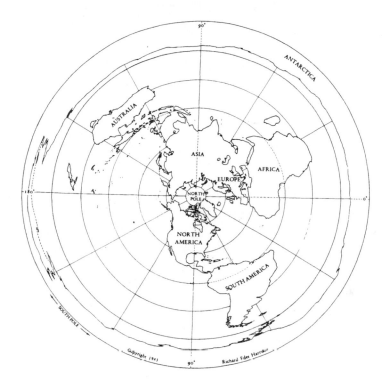

FIGURE 5.6   *Azimuthal equidistant world map centered on the North Pole.*

tember 1987 Montreal Protocol on Substances That Deplete the Ozone Layer was the beneficiary of this widespread awareness, even if it did not, as has been asserted, directly result from it.[51] The influence of "ecospectral mapping" from space will continue. When they are linked to GIS models of environmental processes such as ozone depletion or the greenhouse effect, maps can serve as "crystal balls."[52]

For addressing contemporary international issues of economic and social concern, it is the more conventional equal-area maps that are most frequently called upon. "Distributional" data are best displayed on such projections, for they permit accurate spatial and statistical comparisons to be made. Today it is not only the technical virtues of equal-area maps that are valued for their use in the comparative analysis of data regarding issues such as population growth or mineral allocation or the world distribution of hunger or disease.[53] It is also their emblematic, and therefore political, quality that has made equal-area projections so popular.

In a veritable "battle of the maps," the Peters projection has been pitted against the Mercator projection as the preferred symbol for world order, "a map for our day," in the twenty-first century.[54] Arno Peters himself has charged that it is always the countries of the "Third World," often ex-colonial states, that are "disadvantaged on Mercator's map."[55] His cartographic correction of the alleged Eurocentric bias of Mercator's projection is a map that not only shows countries in correct size relation to each other (at the cost of considerable distortion of shape) but also puts the equator back in the middle of the map (as the modified versions of the Mercator conformal map commonly used today do not). The consequence of these simple changes is to make South America and Africa, as well as Australia, larger and more visually prominent (see Figure 5.2). The Peters map has been enthusiastically taken up by church groups, development institutes, various national governments, and even certain international organizations on the basis of an apparent belief that cartographic equivalence will militate in favor of social and cultural equality, economic equity, and political equilibrium.[56]

Conclusion

The notion of objectivity in cartography is, for the mapmaker, an aspiration and, for the map viewer, a presumption. Objectivity is at the same time an impossible ideal. Not even the globe, which permits us to see the world "round" but not all at once in its entirety, is a perfect map. There is no single framework that will satisfy every criterion of mapmaking, notably both conformality and equivalence, but there are many map concepts and designs that do suffice for particular purposes. The Robinson projection, which is an attempt to circumvent some of the more overt political criticisms of other projections, is described by its maker, Arthur Robinson, as an attempt to "create a portrait," albeit an inaccurate one, of the earth. For Robinson the "artistic" approach came first, which was followed by "the mathematical formula" to produce the visual effect he liked. His map was judged as "the best balance available between geography and aesthetics" by John Garver, chief cartographer of the National Geographic Society, which, in 1988, replaced the Van der Grinten projection with Robinson's projection for its new world map (see Figure 5.7).[57]

Cartography is a combination of science and art, of the objective and the subjective in human thought and activity. Like the "world" itself, the map is both an object and an idea, a material entity and a mental construct. Mapmaking and map viewing are both, therefore, influenced and informed by biases, some obvious and others subliminal. However, a basic knowledge of how maps are planned and how they may be intended, that is, the politics of mapping, enables a map reader to gain the faculty of graphic literacy, and thereby an awareness of the power of maps.

ROBINSON PROJECTION

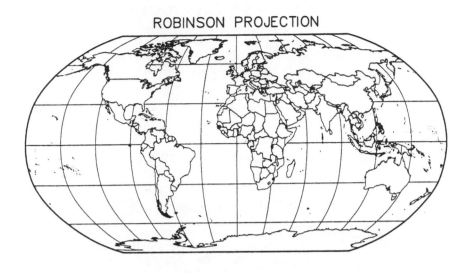

VAN DER GRINTEN PROJECTION

FIGURE 5.7 *A new shape for the world.*

Notes

1. Walter Lippmann, *Public Opinion* (New York: Harcourt, Brace, 1922), pp. 16, 29. Lippmann's reference to "Bohemia" probably was literary as well. In Shakespeare's comedy *The Winter's Tale,* one of the stage settings is BOHEMIA, identified as "A desert Country near the Sea" (Act III, Scene III).

2. Yves Lacoste, "Geopolitics and Foreign Policy," *SAIS Review* 4 (Summer-Fall 1984): 213.

3. President Lyndon B. Johnson sent copies of the *Apollo* photograph to all the world's heads of state, including Ho Chi Minh. Walter A. McDougall, *The Heavens and the Earth: A Political History of the Space Age* (New York: Basic Books, 1985), p. 412.

4. For an extraordinary example of the translation of satellite imagery into finished cartography, juxtaposing computer maps and photographs with conventional thematic maps of the same areas, see *Images of the World: An Atlas of Satellite Imagery and Maps* (Chicago: Rand McNally, 1983). The geographic information system "revolution" is well reported by David Bjerklie, "The Electronic Transformation of Maps," *Technology Review* 92 (April 1989): 54–63.

5. Arthur H. Robinson and Barbara Bartz Petchenik, *The Nature of Maps: Essays Toward Understanding Maps and Mapping* (Chicago: University of Chicago Press, 1976), p. x.

6. S. W. Boggs, "Cartohypnosis," *Department of State Bulletin* 15 (December 22, 1946): 1119–1125. Maps can also be used, as Boggs points out, to "dehypnotize people."

7. J. B. Harley, "Maps, Knowledge, and Power," in *The Iconography of Landscape: Essays on the Symbolic Representation, Design, and Use of Past Environments,* ed. Denis Cosgrove and Stephen Daniels (Cambridge: Cambridge University Press, 1987), pp. 277–312.

8. For informative historical surveys of cartography, see Lloyd A. Brown, *The Story of Maps* (Dover Publications, 1977), Norman J. W. Thrower, *Maps & Man: An Examination of Cartography in Relation to Culture and Civilization* (Englewood Cliffs, NJ: Prentice-Hall, 1972), and John Noble Wilford, *The Mapmakers: The Story of the Great Pioneers in Cartography from Antiquity to the Space Age* (New York: Vintage Books, 1982).

9. A Pawnee star chart from the Field Museum of Natural History in Chicago was included in "The Power of Maps" exhibition at the Cooper-Hewitt National Museum of Design, Smithsonian Institution, New York City, October 6, 1992–March 7, 1993.

10. On this characteristic of Indian mapping, see Gregory H. Nobles, "Straight Lines and Stability: Mapping the Political Order of the Anglo-American Frontier," *Journal of American History* 80 (June 1993): 26–27. Contrast with John Mitchell's "A Map of the British and French Dominions in North America with the Roads, Distances, Limits, and Extent of the Settlements," 1755. A third edition of this map, consulted during the 1783 Paris negotiations to establish the boundaries of an independent United States, is generally regarded as the most important map in American history. Seymour I. Schwartz and Ralph E. Ehrenberg, *The Mapping of America* (New York: Harry N. Abrams, 1980), p. 164.

11. Phillip C. Muehrcke, with the assistance of Juliana O. Muehrcke, *Map Use: Reading, Analysis, and Interpretation* (Madison, WI: JP Publications, 1978), p. 2.

12. Robinson and Petchenik, *The Nature of Maps,* pp. 4–5.

13. For an elaboration, with examples from around the world, see Alan K. Henrikson, "Mental Maps," in *Explaining the History of American Foreign Relations,* ed. Michael J. Hogan and Thomas G. Paterson (Cambridge: Cambridge University Press, 1991), pp. 177–178.

14. This important point is made by Yi-Fu Tuan, "Images and Mental Maps," *Annals of the Association of American Geographers* 65 (June 1975): 205–213.

15. This is persuasively demonstrated by Rudolf Arnheim, *Visual Thinking* (Berkeley: University of California Press, 1969).

16. Among many possible expert discussions of the principles of map projection, see the brief corrective analysis by Mark Monmonier, *How to Lie with Maps* (Chicago: University of Chicago Press, 1991), pp. 8–18, and the more technical treatment of David Greenhood, *Mapping* (Chicago: University of Chicago Press, 1964), chap. 6, "Flat Maps with Round Meanings."

17. Arno Peters, *The Europe-Centred Character of Our Geographical View of the World and Its Correction* (Munich-Solln: Universum Verlag, 1979).

18. Arthur H. Robinson, "Arno Peters and His New Cartography," *American Cartographer* 12, no. 2 (1985): 104. Robinson points out that the "Peters projection"—a cylindrical equal-area map with standard parallels at 45° from the chosen central great circle—was one of three cylindrical-projection variants proposed by the Edinburgh clergyman-cartographer James Gall in 1885.

19. W. R. Tobler, "Medieval Distortions: The Projections of Ancient Maps," *Annals of the Association of American Geographers* 56 (June 1966): 351–360, is a rare effort to consider the technical properties of the Hereford map and other premodern cartography.

20. Brown, *The Story of Maps,* p. 71; see also the Ptolemaic world map on p. 55.

21. Among the realizations that can be gained from viewing such reversed-image maps is, for example, the striking impression that "South" America lies almost entirely east of Florida—and much farther from Asia and closer to Africa—than is generally assumed.

22. See Harrison's essay, "The Geographical Sense," in his *Look at the World: The FORTUNE Atlas for World Strategy* (New York: Alfred A. Knopf, 1944), pp. 10–12, and his remarkable maps themselves.

23. The influence of his work is discussed in Alan K. Henrikson, "The Map as an 'Idea': The Role of Cartographic Imagery During the Second World War," *American Cartographer* 2 (April 1975): 19–53. Harrison's maps were used, for example, to illustrate U.S. War Department, General Staff, *George Marshall's Report: The Winning of the War in Europe and the Pacific;* biennial report of the chief of staff of the United States Army, July 1, 1943, to June 30, 1945, to the Secretary of War (New York: Simon and Schuster, 1945).

24. Monmonier, *How to Lie with Maps,* pp. 89–94.

25. Giselher Wirsing et al., *Der Krieg 1939/41 in Karten* (München: Verlag Knorr & Hirth, 1942).

26. Michael R. Beschloss and Strobe Talbott, *At the Highest Levels: The Inside Story of the End of the Cold War* (Boston: Little, Brown, 1993), pp. 162–163. Cf.

the similar arguments of an earlier Soviet leader, Leonid Brezhnev, in Henrikson, "Mental Maps," pp. 182–184.

27. Monmonier, *How to Lie with Maps,* pp. 19–20.

28. Ibid., chap. 10, "Color: Attraction and Distraction." Monmonier emphasizes that the "decorative" role of color—in television cartography, for example—can easily interfere with its "functional" role.

29. Rudolf Arnheim, "The Perception of Maps," *American Cartographer* 3 (April 1976): 6.

30. The term is that of Christian von Ehrenfels. Wolfgang Köhler, *Gestalt Psychology* (New York: Mentor Book, 1947), pp. 102–105.

31. "Follow me . . . I am the earth in the palm of your hand," *Smithsonian* 23 (February 1993): 112–117. See also Denis Wood, *The Power of Maps* (New York: Guilford Press, 1992).

32. That the federal government is, by far, the preeminent patron of mapmaking in the United States is implicitly demonstrated in Ralph E. Ehrenberg, *Scholars' Guide to Washington, D.C., for Cartography and Remote Sensing Imagery (Maps, Charts, Aerial Photographs, Satellite Images, Cartographic Literature and Geographic Information Systems)* (Washington, D.C.: Woodrow Wilson International Center for Scholars, Smithsonian Institution Press, 1987).

33. Harley, "Maps, Knowledge, and Power," p. 281.

34. A story by Jorge Luis Borges imagines a map, produced by a government's College of Cartographers, that is physically coextensive with the country! Jorge Luis Borges, "Of Exactitude in Science," excerpted in Nobles, "Straight Lines and Stability," p. 9. For the expression of geography in prose and poetry, see also Jorge Luis Borges, in collaboration with María Kodama, *Atlas* (New York: E. P. Dutton, 1985).

35. On the iconographic setting of Washington, D.C., in national and international context, see Alan K. Henrikson, "'A Small, Cozy Town, Global in Scope': Washington, D.C.," *Ekistics: The Science and Study of Human Settlements* 50 (March-April 1983): 123–145, 149.

36. J. B. Harley, "Silences and Secrecy: The Hidden Agenda of Cartography in Early Modern Europe," *Imago Mundi* 40 (1988): 65, 66.

37. Harley, "Maps, Knowledge, and Power," pp. 302–303.

38. On Jefferson, see William H. Goetzmann, *New Lands, New Men: America and the Second Great Age of Discovery* (New York: Viking, 1986), pp. 110–118. Meriwether Lewis and William Clark, "A Map of Part of the Continent of North America" (1809), is reproduced on p. 116. On westward movement and the role of mapping, see also *idem, Army Exploration of the American West, 1803–1863* (New Haven, CT: Yale University Press, 1957) and *Exploration and Empire: The Explorer and the Scientist in the Winning of the American West* (New York: Alfred A. Knopf, 1967). On Americans' geographical orientation toward the Pacific more generally, see Norman A. Graebner, *Empire on the Pacific: A Study in American Continental Expansion* (New York: Ronald Press, 1955), and Frederick Merk, *Manifest Destiny and Mission in American History: A Reinterpretation* (New York: Vintage Books, 1963).

39. Among the earliest U.S.-centered world maps were David H. Burr's "The World, on Mercator's Projection," published by J. Haven in Boston in 1850, and

"Colton's New Illustrated Map of the World on Mercator's Projection," published by J. Colton in New York in 1851. Curiously, an even earlier German map placed North and South America at the world's center. "PLANIGLOB in Mercator's Projection. Zugleich als KARTE v. AUSTRALIEN," map II.b in Adolf Stieler, *Schul-Atlas* (Gotha: Justus Perthes, 1841). This information was obtained by courtesy of John A. Wolter and Richard W. Stephenson, Geography and Map Division, Library of Congress.

40. On this theme of U.S. power and cartographic centrality, see Alan K. Henrikson, "America's Changing Place in the World: From 'Periphery' to 'Centre'?" in *Centre and Periphery: Spatial Variation in Politics,* ed. Jean Gottmann (Beverly Hills, CA: SAGE, 1980), pp. 79–80, 95n; the Mitchell *Atlas* world map is compactly reproduced on page 80.

41. The standard historical interpretation is Dexter Perkins, *A History of the Monroe Doctrine,* rev. ed. (Boston: Houghton Mifflin, 1963).

42. The use of the double-hemisphere world map increased in popularity after such a map appeared in Mercator's *Atlas sive cosmographicae* (Duisburg, 1595). The usual projection for double-hemisphere maps is the orthographic, which shows half the earth as if viewed from an infinite distance (see Figure 5.1).

43. S. W. Boggs, "This Hemisphere," *Department of State Bulletin* 12 (May 6, 1945): 845–850.

44. Henrikson, "The Map as an 'Idea,'" pp. 19–20, 28–33; Arthur P. Whitaker, *The Western Hemisphere Idea: Its Rise and Decline* (Ithaca: Cornell University Press, 1954).

45. On the genesis and the diplomacy of NAFTA, see Alan K. Henrikson, "A North American Community: 'From the Yukon to the Yucatan,'" in *The Diplomatic Record, 1991–1992,* ed. Hans Binnendijk and Mary Locke (Boulder: Westview Press, 1993), pp. 69–95.

46. Robert M. Lunny, *Early Maps of North America* (Newark, NJ: New Jersey Historical Society, 1961). One of these older maps of North America before political partitioning was used to publicize "The North American Concept: A Symposium on Issues That Affect Canada, the United States, and Mexico, and How They Are Internationally Managed," Center for the Study of Foreign Affairs, Foreign Service Institute, U.S. Department of State, April 15, 1987.

47. As early as 1941, the economist Eugene Staley, in "The Myth of Continents," *Foreign Affairs* 19 (April 1941): 481–494, attacked the tendency toward regionally based economic groups by pointing out that, in terms of cost-distance, cities in the United States were sometimes "closer" to cities in Europe or even in Asia than they were to each other.

48. Henrikson, "The Map as an 'Idea,'" pp. 45, 53n.

49. Alan K. Henrikson, "'Wings for Peace': Open Skies and Transpolar Civil Aviation," in *Vulnerable Arctic: Need for an Alternative Orientation,* ed. Jyrki Käkönen (Tampere, Finland: Tampere Peace Research Institute, 1992), pp. 107–143.

50. For example, the map of ozone levels in the Southern Hemisphere based on data from the Total Ozone Mapping Spectrometer (TOMS) on board the National Aeronautics and Space Administration's *Nimbus 7* satellite. Richard S. Stolarski, "The Antarctic Ozone Hole," *Scientific American* 258 (January 1988): 30–36; map on p. 31.

51. Richard Elliot Benedick, *Ozone Diplomacy: New Directions in Safeguarding the Planet* (Cambridge: Harvard University Press, 1991), pp. 18–20, 110–111.

52. Bjerklie, "The Electronic Transformation of Maps," p. 61.

53. A classic study of the maldistribution of food in the world is Josué de Castro, *The Geography of Hunger* (Boston: Little, Brown, 1952). A recent work emphasizing the spatial misallocation of public health efforts is Peter Gould, *The Slow Plague: A Geography of the AIDS Pandemic* (Cambridge, MA: Blackwell, 1993).

54. Scott Minerbrook, "'Mental Maps': The Politics of Cartography," *U.S. News & World Report* 110 (April 15, 1991): 60.

55. Peters, "The Europe-Centred Character of Our Geographical View of the World and Its Correction," p. 7.

56. One of the first prominent usages of the Peters map was its appearance on the front cover of the report of the Independent Commission on International Development Issues, chaired by the Social Democratic former West German Chancellor Willy Brandt: *North-South: A Program for Survival* (Cambridge: MIT Press, 1980). On this map an awkward curved line is drawn, crossing the Equator, to try to differentiate the "North" from the "South" more clearly than the Peters projection itself can do. Australia, which by virtue of its very name is southern, is delineated as being within the "North."

57. Bruce Van Voorst, "The New Shape of the World," *Time* 132 (November 7, 1988): 127; John Noble Wilford, "The Impossible Quest for the Perfect Map," *New York Times,* October 25, 1988. The Robinson map is based on an ellipse, with the lines of longitude curving toward the poles; the latitude lines are straight. His standard parallels are 38° north and 38° south. Only at these latitudes are size and shape relationships accurate, as on a globe.

Chapter Six

Electoral Geography and Gerrymandering: Space and Politics

RICHARD MORRILL

For those societies in which elected representatives have meaningful power and in which they are elected from districts, the geographic design of districts is itself a major element of the balance of power. This chapter briefly reviews how manipulation of the layout of electoral districts has been used to influence the distribution of power. Geographers are interested in territorial behavior—how and why people organize themselves on the landscape—as through electoral districts.

The emphasis here is on the United States experience, but the principles, if not the details, are rather universally applicable. A short historical review of the idea of territorial representation, from the battle over malapportionment to racial discrimination to gerrymandering follows. This is supplemented by a discussion of criteria for and methods of redistricting, and I conclude with a review of recent trends and critical issues, particularly the sacrifice of geographic integrity in redistricting efforts.

International Variation in Electoral Systems

The United States system of election by plurality or "winner take all" within individual districts of constituencies is characteristic of relatively few countries—Canada, the United Kingdom, Australia, New Zealand,

South Africa, Chile, Japan, and France. The system diffused from Britain to many of its former colonies (the United States imposed Japan's postwar constitution). The Australian senate employs the "alternate vote" system to correct somewhat for the electoral bias prevalent in these systems. France requires a double ballot or runoff election to ensure that the winner obtains a majority. Japan elects 511 Diet members from 124 districts; since each voter gets only one vote, some of the bias can be overcome.

Most countries of the world, however follow various forms of proportional representation to ensure that parties (over some minimum threshold) receive seats in proportion to their share of popular votes. Ireland employs a quite complex system called the "single transferable vote." Germany employs a rather simple and ingenious system called "additional member"; each voter votes twice: in a constituency, as in the United States and United Kingdom, but also for a party within a state (Lander). The latter is used to "top up" or achieve proportional representation overall, clearly reducing the incentive to gerrymander. Nevertheless, territorial manipulation may be useful to protect individual incumbents or representation for particular places.

The large majority of countries in northern and southern Europe, Latin America, and Asia employ various forms of a "list" for proportional representation. Voters usually do not vote for individuals but for party slates within large multimember districts, which are in fact usually traditional administrative areas. The extreme cases of proportional representation are Israel and the Netherlands, where the entire nation is one constituency. In the countries with proportional representation, electoral districts may be unimportant, but changes in definitions of administrative territories, for example, boundaries of major cities, can be quite important, since it greatly influences the variation in partisan control and the election of important regional figures (governors, mayors).

Proportional representation is often advocated by the left (as the British Labour Party) on the grounds that their supporters are more concentrated in large cities. However, the Democrats in the United States vastly benefit from the individual district system, since they are able to dominate many districts with very small voter turnout, which would be submerged by areawide Republican majorities.

There are unusual examples of electoral systems in which representation may be granted to groups other than in bounded, territorial districts or political parties. The system used at the end of the Soviet period in Russia granted a third of all seats in the People's Congress of Deputies to special interest groups. Thus a third of the members were elected from a set of organizations such as the labor unions, Academy of Sciences, All-Union Geographical Society, and many other such groups.

Malapportionment, Racial Discrimination, and Gerrymandering: The Constitutional Road

Today the electoral district is quite ephemeral, but until this century it did not differ from the basic political subdivisions of countries. Thus, traditionally each English borough was entitled to two seats in the commons. The English and Scandinavian idea that representation was territorial and not strictly according to "who you were" was fundamental. After all, representation might have been accorded by social position or interest group or occupation. Most democratic societies adopted a territorial basis for representation to reflect the actual regional structure that had evolved historically—based on the landed aristocracy, the rise of cities and exchange, and cultural variation. Representation on the basis of these traditional units gradually departed from a sense of their having an approximately equal stake in the nation. As the franchise extended to all adult males, and as the population became more urban, the issue of fairness of the system of districts arose.

The manipulation of territory for political ends is hardly new. Complaints about gross inequality in population were already being lodged in eighteenth-century Britain against the "rotten" boroughs. Malapportionment, or inequality in the population size of districts, has been of concern for centuries, and gerrymandering, or the manipulation of the boundary and composition of districts for partisan purposes, for almost as long.

Many states chose to maintain representation from traditional territories by means of proportional representation—a province receives seats in relation to its share of the national population, and within the area, parties receive seats in relation to their share of the total vote. The United States Senate continues pure representation by territory—two senators per state. But in Britain and most of its former colonies, the lower house members are elected from single member, winner-take-all districts, and there has been a gradual acceptance of the principle of somewhat equal population per district. Over time, it has been more and more difficult to create districts that are composed of any traditional administrative or cultural areas.

The United States Constitution provides for the allocation of representatives among states and guarantees that each state will also have a representative system of government. Indeed, the very existence and continuity of the census is based on the constitutional mandate to reapportion seats in the House of Representatives among the states. With the total number of seats remaining the same, if the population of the states changed, then so did their allotment, thereby requiring redistricting within those states. Even so, until the 1970s, it was customary, except in the very largest cities, for both congressional and state legislative districts to be composed of whole counties or cities.

Gross levels of malapportionment within states and degrees of gerrymandering were routinely criticized throughout the nineteenth and twentieth centuries in the United States. But as recently as 1946, in *Colgrove* v. *Green,* the U.S. Supreme Court dismissed a complaint that a Chicago congressional district with 915,000 people was treated unfairly compared to a rural one with 112,000. Most states had long ignored their own constitutional requirements for redistricting and routinely overrepresented rural areas; one-third of the states maintained the "Little Federal" scheme, with state senates having one representative per county, regardless of population size.

The Reapportionment Revolution

By 1960 the underrepresentation of urban areas had become extreme. In 1962, in *Baker* v. *Carr,* the U.S. Supreme Court established the general principle of population equality among congressional districts and introduced the phrase "one man, one vote" in striking down in 1964 Georgia's congressional plan that included an Atlanta district of 824,000 and a rural district with 272,000. In the same year, in *Reynolds* v. *Sims,* the Court extended the principle of substantial population equality to both houses of state legislatures, stating that "people, not trees, vote." By 1970, the principle had been extended downward to all representative bodies, including county commissioners, city councils, and school districts. And by 1975, and via a long series of court cases, all fifty states had redistricted both their congressional and legislative districts. As a result, malapportionment effectively ceased to be a major issue.

Gerrymandering, the purposeful drawing of districts to benefit some and restrict others, remained. The major purposes of such discrimination are related to political parties, race or ethnicity, incumbent members, and kind of territory (e.g., urban versus rural). It surprises many that except for gerrymandering against a racial or ethnic minority, gerrymandering remains not only legal but normal and widespread. The term "gerrymander" immortalizes Governor Gerry of Massachusetts, who signed into law a districting designed to maximize the election of Republican-Democrats over Federalists. A classic example of gerrymandering is shown in Figure 6.1, where the layout of districts was blatantly designed to prevent the election of a black congressman, after federal marshals had begun to register blacks to vote. In Illinois, New York, and Ohio, among others, gerrymandering long maintained Republican domination of the legislatures, despite statewide Democratic majorities. In recent years, Democrats used the same techniques to try to maintain dominance in the face of statewide Republican majorities.

FIGURE 6.1   *The original gerrymander (from the* Boston Gazette, *March 26, 1812).*

SOURCE: Richard Morrill, "Political Redistricting and Geographic Theory," *Resource Publication in Geography* (Washington D.C.: American Association of Geographers, 1981).

Racial Gerrymandering

Even after districts became more equal in population, and as blacks began to register and vote, legislatures tried to minimize black representation. In 1966, the Supreme Court held that states could not draw districts with the intent of minimizing the possibility of blacks getting elected. After the Voting Rights Act of 1968, the courts have gradually become more and more restrictive, not only striking down plans with any appearance of discrimination but virtually requiring what is called "affirmative gerrymandering" in favor of the chances of minority election. Interestingly, in June 1993, the U.S. Supreme Court ruled that such "gerrymandered districts" can violate the rights of white voters, casting some doubt on the validity of at least extreme cases. More recent court rulings have further complicated this issue and it has not been fully resolved. The protected groups under the Voting Rights Act are racial and linguistic minorities—the latter primarily people of Hispanic origin. The Court has also disallowed the use of multimember districts, where it was shown that the purpose or intent was to dilute minority representation.

Within the past few years, there have been several cases aimed at partisan political gerrymandering. *Bandemer v. Davis* (1986) concerned an obvious Republican gerrymander of the Indiana legislature, relying on multimember urban districts to dilute Democratic chances. *Badham v. Eu* (1989) dealt with the even more blatant and quite effective Democratic gerrymander of the congressional districts of California. Nevertheless, in both cases the Supreme Court held that the other party could conceivably have won, despite the gerrymandering, and thus decided that there was no constitutional issue of electoral fairness—that is, some persons' votes were not made less "effective."

Criteria for and Methods of Redistricting

The purpose of voting is to enable people to express their will with respect to issues of collective choice.[1] One may vote directly on issues, or indirectly via supporting a party that expresses a program with which one agrees, or via a representative who shares a sense of belonging to a territory and reflects its interest, or via a representative who shares a racial or linguistic heritage that may have been historically suppressed. The goal of districting is to make possible the meaningful participation of voters in electing individuals who meet these senses of representation: party, place, and perhaps race. A voter needs to feel that a vote matters. There can be better or poorer districting. If done poorly, districting can create a sense of disenfranchisement and futility. Characteristics of poor quality districting are as follows:

1. malapportionment;
2. fragmentation of the territorial base of a party so that it cannot, over time, win seats in proportion to its popular appeal;
3. overconcentration of a party's adherents so that its strength is wasted and it cannot win seats in proportion to its popular appeal;
4. manipulation of territory to unfairly advantage or disadvantage incumbents of particular parties;

5. fragmentation of the territorial base of a racial or linguistic minority;
6. overconcentration of a racial or linguistic minority;
7. unnecessary fragmentation or division of a territory with which people identify, or of districts with which voters traditionally identify;
8. very high proportion of very safe seats so that voters in general feel that there is no possibility of change;
9. very high proportion of highly competitive seats so that representatives are not elected long enough to gain experience or commitment.

Characteristics 2 through 9 are all forms of gerrymandering, with respect to party, race, language, or meaningful territory. Pre-election evidence of poor districting can be gleaned from analysis of prior election returns and maps. Outcome evidence of the effectiveness of gerrymandering can be obtained on the basis of who runs or retires and from turnout, as well as from electoral returns.

Constitutional Criteria for Districting Equal Population

Equal Population.   The degree of population equality depends on the level of jurisdiction. Congressional districts must be well within one percent of the average population of a state's congressional district—that is, about 5,000 people—for the 1990s redistrictings. Legislatures, county commissioners, and the like, are given much greater latitude, often plus or minus 10 percent, especially if the states are small and population per district is small. States with large populations per district, such as California, tend to require fairly strict equality. Population experts argue that extreme population equality—often carried to within a few persons—is absurd, given the uncertainty of the census count, the likelihood of undercount, and the rapid change in population during the years between censuses. In practice, a decision on population equality may give the court a backhand excuse to strike down a grossly gerrymandered plan (as the New Jersey congressional plan in 1981) without ruling on the gerrymander as such.

The United States appears to be far more extreme regarding the criterion of equal population than other countries, including Canada and the United Kingdom, where traditional arguments prevail that rural areas have greater need for representation.

Racial Fairness.   The Fourteenth and Fifteenth Amendments to the Constitution and the Voting Rights Act of 1965, 1968, and 1982 have created a strong requirement for equal representation or fairness for racial and linguistic minorities. The VRA lays out criteria defining racial discrimination and procedure for judicial review of districting plans, taking into account any history of discrimination, minority access to candidate selection, other forms of discrimination in the region, as well as the composition and design of districts.

With modern computers and programs, plans with equal populations are very easy to create and the main basis for judicial complaint is based on grounds of discrimination against racial or linguistic minorities. In effect, a criterion has evolved that requires the redistricting entity—usually a legislature or other governing body but at times courts or special commissions—to maximize the number of districts that would permit the likely election of a minority. This usually means that districts be drawn to have at least a 50 percent voting-age population; for Hispanics in most areas, and for blacks in many areas, this also means a 60 to 65 percent majority of the total population. In order to achieve this, gerrymandering, or the stringing together of appropriate territory is expected and encouraged, at least until the June 1993 Supreme Court ruling on racial gerrymandering.

Other Criteria

One may be surprised that the only required criteria are those cited above. Some state constitutions add other criteria, but their weight is inconsequential compared to the above.

Compactness. Geographers know not to expect too much from a compactness criterion. One can discriminate and maintain appearances! Compactness is rather an operational aid in avoiding discriminatory gerrymandering. Compactness is preferable to irregularity simply because compact territories make for easier communication and internal cohesion. What is of concern are egregious irregularities that reveal an intent to discriminate. Much work has been spent on measures of compactness. Yet the courts have not been impressed. The better measures are based on population distribution rather than external boundaries, for example, the proportion of population inside or outside the district within a polygon enclosing the district.

Respect for Integrity of Political Units. This criterion is and should be an important one. Local governments are vital, legal, and familiar communities of interest since in much of the country, counties have been a traditional basis of representation, the building blocks for larger districts, and represent a simple barrier to extreme gerrymandering for racial, partisan, or geographic purposes. It is a criterion of wider purpose than compactness and deserves higher priority. The courts generally agree.

Community of Interest. This is the least well defined but the most geographic criterion, in the sense that a major concern of geography is to identify the territories with which citizens identify. The idea of community of interest overlaps with both racial and partisan identity and with political jurisdictions. Communities are revealed through patterns of work, of residence, and of social, religious, and political participation. At a broad scale, there is divergence of identity between an urban core, suburbs, and rural and small town areas because they are different jurisdictions, have different needs, and attract people with different values. There are also those regions defined by a small city and its hinterland. Within a metropolis are found broad areas of similar age, class, and political leaning.

Why Does Community of Interest Matter?   As stated above, one of the three bases of representation is territorial—not of arbitrary aggregations of geography or space for the purpose of conducting elections but as meaningful entities that have legitimate collective interests arising from the identity of citizens with real places and areas. Since it is the people of real places that pay taxes to support national and local activities, it is not unreasonable that they should develop a sense of common interest as to how the money is spent. It was because of a profound sense that territorial interests (that is, cities) were not being fairly represented that led to the reapportionment revolution in the first place. No matter in how partisan a manner we may vote, the representative also stands for the district. This is the very essence of our federal-territorial system of governance and why the United States does not have proportional representation. It is also why gerrymandering matters.

District Stability.   If it is true that voters develop over time a sense of cohesion, then the district becomes "real," as exemplified by organizations, meetings, and the like. A defensive case can be made for minimizing the number of voters who would have to be shifted to another district and whose customary allegiance would be disrupted. However, this has the simultaneous effect of protecting incumbents and of decreasing the number of competitive seats and the responsiveness of the system to swings in voter sentiment. Also, if an existing set of districts were very poorly drawn, there would be no virtue in preserving its imperfections.

Electoral Fairness.   The principle of electoral fairness reflects the idea of a "level playing field"—that over a set of districts, over a period of time, the number of seats won will be in proportion to the share of votes. Avoidance of discriminatory packing and splitting of partisan concentrations is not totally avoidable because of the geographic concentrations of partisan strength. The point is that the parties should be treated symmetrically in the redistricting process—that one party is not disproportionately favored or hurt. When such discrimination occurs, it can be measured through the use of registration data and returns from recent elections. Obviously, the parties use such data as tools to gerrymander effectively.

Why is this criterion important, despite the Supreme Court's indifference? Why should parties be entitled to protection? Such questions cut to the very core of the theory of representative democracy—a theory based on dual premises: the will of the majority and respect for the rights of the minority. Fairness implies that majority parties should win legislative majorities whereas minority parties should nevertheless win seats in proportion to their numbers. It is not the parties but the views of voters that matter. Voters choose representatives, and therefore parties, in order to entrust them with governing. In a winner-take-all system, the voters accept the reality that the other party may win. But the very idea of fairness and will of the majority insist that over the long run, there will be a fair correspondence between the share of votes received and seats won. Otherwise the frustrated minority can only conclude that they are effectively disenfranchised. How blatant and destructive does gerrymandering have to be before fairness tells us that it should be invalidated? The presence of some electoral bias—a difference between the share of votes and seats—is not enough since some disparity may be the result of the natural distribution of partisan strength, such as relatively "safe" areas for one or the other party. Rather there needs to be a linked set of indicators: electoral bias; the presence before the election of a discriminatory opportunity for election; asymmetry in the distribution of partisan majorities; splitting and packing or incumbent discrimination; irregularity and break up of communities of interest and of political units; a process that excludes the minority party; and lack of competitiveness.

Partisan gerrymandering is not always successful, and its effects can be transient. This occurs when the perpetrators have overreached, were overconfident about a supposed voter realignment, and while packing the other party, riskily tried to maximize their number of seats with smaller margins. When more adroitly done, gerrymandering can be successful. Otherwise, gerrymandering would not have been so common over the generations and a subject of continuing debate and litigation.

Balance of Safe and Competitive Seats.   This is a difficult and controversial criterion. The idea is that each party should have a similar share of fairly safe seats (for example, won by more than 60 percent of the vote) and of competitive seats and that, overall, probably no more than two-thirds of seats will be safe. Given the 95 to 99 percent reelection success of congressional incumbents in the United States, some have proposed a higher proportion of competitive seats and/or a limitation on the number of terms. The argument for more competitive seats reflects a desire for greater elasticity or responsiveness to swings in voter sentiment; the argument for keep-

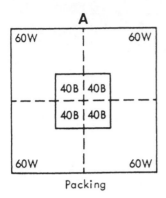

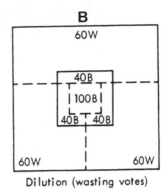

FIGURE 6.2   *Racial gerrymandering.*

SOURCE: Richard Morrill, "Political Redistricting and Geographic Theory," *Resource Publication in Geography* (Washington D.C.: American Association of Geographers, 1981).

ing safe seats values continuity of leadership, accumulation of experience, and balance of legislative versus executive power.

Methods of Gerrymandering and of Redistricting

There are four useful techniques for gerrymandering defined as manipulating territory for unequal power:

1. Packing or wasting the opposition
2. Splitting or diluting the opposition
3. Differential treatment of incumbents
4. Selective use of multimember districts

If the minority—whether a political party, racial or ethnic group, or urban or rural territory—is much smaller than the majority, then a degree of "affirmative gerrymandering" or manipulation may be practiced in order to approach parity of votes and seats.

Packing is illustrated in Figure 6.2A. It is simply a matter of overconcentrating the "minority" so that their votes are "wasted"; they win too few seats by too large margins. Thus ghetto blacks in Chicago or New York were afforded one district that was 90 percent black, when their numbers were sufficient to elect two or more representatives. Packing occurs when it is impossible to deny the other party some representation. Beyond that core, or where minority concentrations are not so large, additional, or even any representation, can be denied by the practice of diluting the minority

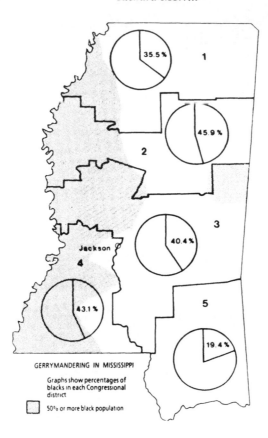

FIGURE 6.3   *Gerrymandering of congressional seats in Mississippi. As Mississippi blacks became registered, congressional districts were shifted from a north/south to an east/west pattern, preventing any district from having a black majority population.*

SOURCE: Richard Morrill, "Political Redistricting and Geographic Theory," *Resource Publication in Geography* (Washington D.C.: American Association of Geographers, 1981).

vote. This is illustrated in Figure 6.2B, where a concentration large enough to elect one representative is split among several districts, each with a majority for the gerrymandering party. This was the technique used by the Mississippi legislature to prevent election of a black representative from Mississippi, despite the fact that blacks constituted 39 percent of the state population. As a result of court litigation, the Mississippi districts were redrawn, resulting in the election of the first rural black representative since the Reconstruction period after the Civil War (see Figure 6.3).

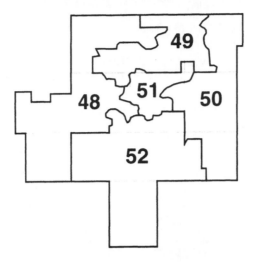

FIGURE 6.4   Multimember districts around Indianapolis–Marion County.

SOURCE: Richard Morrill, "Political Redistricting and Geographic Theory," *Resource Publication in Geography* (Washington D.C.: American Association of Geographers, 1981).

Because of the great value of incumbency, discriminatory treatment can be very effective. The idea is to place members of the same party in the same re-drawn district, while creating open districts without incumbents but with a majority of voters for the gerrymandering party. In the famous Burton redistricting of California congressional districts after 1980, the techniques of packing, dilution, and discrimination against incumbents were all used. As a result, Democrats received six more seats than their share of the votes would imply; half of these could be attributed to discrimination against Republican incumbents, and half to a mixture of packing and dilution.

The fourth technique is to create multimember districts with a moderate but safe margin for the gerrymandering party. This is well illustrated by the equally famous Republican Indiana legislative redistricting (see Figure 6.4) in which the large Democratic vote in greater Indianapolis was submerged by yet a larger suburban Republican vote. It is a sad commentary on the role of race in the United States that the gerrymandering was probably made possible only by the collusion of some Democrats who feared the control of the Indianapolis Democratic establishment by blacks; recall that the so-called good government merger of Indianapolis and Marion County was in fact a scheme for Republican control of Indianapolis and for prevention of a possible black mayor.

Does Gerrymandering Work?

From the above examples, gerrymandering obviously does work. In an evaluation[2] of all the congressional districting of the 1980s, electoral bias or difference between seats won and votes received were compared for states that had partisan, bipartisan, or court or commission redistricting (as well as adherence to other criteria such as communities of interest, compactness, and integrity of political units). States with intentional or "strong" partisan redistricting stood out as having quite successfully raised the share of seats 11 percent above the share of votes. Competitiveness was very low, and in the worst cases, California, New Jersey, Indiana, Pennsylvania, and Massachusetts, political units and communities of interest were routinely split with classic gerrymandering irregularities perpetrated. States with bipartisan redistricting—that is, a split in power between the legislature and governor or between houses of the legislature—had low electoral bias but also low competitiveness, since incumbent protection loomed so large. Commission and court redistricting were, on average, superior to partisan or bipartisan redistricting across all the measures, especially at increasing competitiveness. Size of states mattered independently; in general, the smaller the states, the better the redistricting, perhaps because there is less "room" for manipulation. Closely balanced states were worse than "one-party" states, perhaps because each party, when able, is tempted to employ drastic measures to maintain or enhance their fragile power. In less balanced states, gerrymandering was more often tied to ideological or urban versus rural than to strictly partisan interests.

Redistricting Techniques and Models

An important consequence of the reapportionment revolution was the development of technology for redistricting. The greater need to redistrict also coincided with the revolution in computers, computer software, and computer graphics. Already in the 1960s, political scientists, geographers, and others began developing relevant models. Two major approaches are what might be called accounting/graphics models and optimizing models. The first began with a simple technology of assigning subunits to districts, such as census tracts or precincts, and calculating their population and political profiles. This has evolved to a high level of sophistication, so that for the 1990s redistricting, a number of national firms provided graphics redistricting GIS software packages, which permit the user to select pieces of territory on the screen and obtain instant district profiles and then to trade bits of territory among districts to refine the sets of districts. This capacity is immensely popular with politicians as well as with all interest groups.

The full technology was possible only for the 1990s round of redistricting since it was based on the Census Bureau's completion of the TIGER files of census geography at the block level for the entire United States, to which could be attached population and political data.

Optimizing models were developed in the late 1960s and 1970s. These are of the location-allocation type in which the objective function is to minimize total travel (distance) of voters to hypothetical centers of districts, subject to some constraints—equal population and territorial integrity. This type is a size-constrained, optimum location model. An example is the LAP algorithm developed by Massam and Goodchild. Since the objective function assures compactness, such models would seem ideal for redistricting. Indeed, they work quite well and have been used by commissions in Iowa and Connecticut. They are not, however, popular with political redistrictors, who do not view compactness as particularly important, because the models are unable to directly incorporate political criteria. Compactness can, where one of the parties is more concentrated than the other, result in electoral bias against the more concentrated. Thus a simple compactness criterion will tend to pack Democrats unfairly in northern cities and Republicans unfairly in southern cities. Also if communities are more linear than circular, as along a river valley or sector of a city, the model may be too simple and violate their integrity. A simple model places too much weight on compactness and not enough on race, community, and political fairness. As a result, these models are more likely to be used to develop a set of alternative "first-cut" configurations, which are subsequently refined politically. Clearly, to the extent that racial sensitivity requires affirmative gerrymandering, the traditional models are inadequate. It is possible to modify such models, for example, to minimize "social distance" rather than geographic distance and therefore to incorporate race and class and political allegiance. Some experiments along these lines are reported elsewhere.[3]

Recent Developments and Continuing Issues

In the United States, the recent round of massive redistricting of congressional and legislative districts has been completed. The process was rather quick because of the availability of the census returns, because of the existence of the TIGER geography file, and because of the prior preparation of legislatures, interest groups, and specialized redistricting consultants. Four issues deserve a little more discussion: first, the matter of undercount and census adjustment; second, the advantages and disadvantages of computer technology; third, the question of partisan gerrymandering; and fourth, the problem of geographic integrity and racial affirmative gerrymandering.

Census Undercount and Adjustment

No census is complete. In a country as large, diverse, and individualistic as the United States it is astounding that some 97 to 98 percent of the population was counted. Nevertheless, the 2 to 3 percent not counted were not evenly distributed geographically. The undercount was greatest in areas of racial and ethnic minorities, especially Native Americans, blacks, and Hispanics, and in inner-city poverty areas with transient populations. An undercount of 10 percent or more may have occurred in such areas. On the other hand there was probably an overcount in some other areas, especially of the elderly, or where college students were counted at home and at school. The result was to shift political power from the poorer and weaker to the richer and stronger, from the largest cities, such as Los Angeles and New York, to suburban and rural small town regions. The magnitude and geographic concentration of the undercount was sufficient to imply the reallocation of several congressional seats among the states and similar redistribution within some states.

Although Census Bureau statisticians and demographers demonstrated that they could adjust the counts to be more accurate at higher levels of geography—states, counties, and cities, they could not do so at the level of the block, which is critical to the contemporary redistricting process. As a result of both these technical issues and the obvious politics of those who would benefit from adjustment, the Bush administration decided not to adjust the census.

By 1997, the Census Bureau determined that it could, should, and would adjust the 2000 census results for the likely undercount. Republicans in Congress appeared determined to forbid the use of "sampling," in effect outlawing adjustment, as they feared adjustment would shift seats to large city Democratic areas. The basis for adjustment is from intensive postenumeration surveys of undercounted areas and demographic analysis, comparing counts with expected numbers from birth and death records. Setting aside the politics, a vital geographic and statistical issue is at the heart of the controversy: adjustment would probably improve the numbers for units with 5,000 or more persons. However, adjustment—that is, adding or subtracting fully configured households and persons at the block level—would entail an unacceptable level of risk: the failure to recognize that blocks are real places with real people, not just units for assembly into larger districts. It is uncertain whether adjustment could be confined to higher levels of geography.

Advantages and Disadvantages of Computer Technology

In two consulting exercises for Illinois and for California, I had the opportunity to work with the most sophisticated graphics techniques for redistricting. It was possible to develop and experiment with alternatives amazingly quickly. A further advantage was that many citizens groups, for example, racial minorities, were now able to develop their own plans or to evaluate the plans of others. On the other hand, a serious problem was that redistricting had become an exercise in micro-adjustment and territorial adjustment, or reconciling lots of special interests at the margin, so that it became difficult to make plans that had reasonable macro-level validity that meant anything to the residents. Another problem was the encouragement of users to push the equal population requirement to the point of absurdity, shifting blocks about to get mathematical precision, no matter what it did to people and places. The availability of population data at the block level, even if limited, has been a boon to planners and researchers but, together with GIS, has been the impetus for the worst excesses in gerrymandering and manipulation of territory. Possibly the only cure to the dilemmas of census adjustment and of misuse of GIS capabilities is to give up on population data at the block level.

Partisan Gerrymandering

Evaluation of the 1990s round of redistricting suggests that the magnitude and severity of gerrymandering, for both partisan and territorial discrimination, are undiminished. In general this means further erosion of the principles of community, of interest, and of geographic integrity toward a purely functionalist view of electoral districts. Although it is also true that consistent partisan allegiance probably is becoming ever weaker and that therefore the efforts at gerrymandering are really an exercise in political futility, this does not address the erosion of the meaningfulness of districts to voters.

Affirmative Gerrymandering and Geographic Integrity

The most dramatic development of the past few years has been the rise of affirmative gerrymandering for racial and linguistic minority representation. The Voting Rights Act requires that redistrictors aim at achieving proportionality, that is, a share of seats approaching the minority's share of the population. This has come to be interpreted by almost all involved to mean maximizing the number of districts with a greater than 50 percent voting-age population, however much gerrymandering is required, on the grounds that it is never possible to reach proportionality. The maximum number of black, Hispanic, and other racial or general minority districts must be carved out first, after which the remaining districts can be delineated. The effect, in about half of all states with large minority populations, has been to ensure that probably a majority of all districts are determined by the sin-

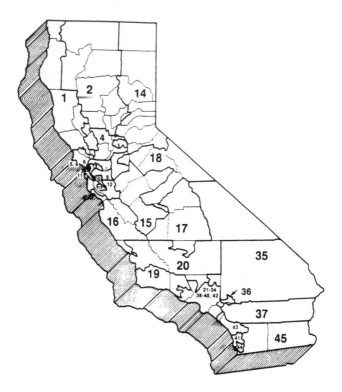

FIGURE 6.5 *California congressional districts, 1980. California had the most infamous gerrymanders.*

SOURCE: *Congressional Districts, 1980* (Washington, D.C.: Congressional Quarterly, Inc., 1980).

gle criterion of race or ethnicity (and population) and that any other sense of community or political integrity is sacrificed. Again, as noted above, this will be affected by the 1993 Supreme Court and subsequent rulings.

This development creates a dilemma for the geographer who is very concerned with racial discrimination but who is also dedicated to meaningfulness of electoral districts. For example, a consultant prepared a plan for Chicago, Illinois, that was astounding for its irregularity but that created eighteen minority districts. It was argued that the correct proportional share was only fifteen and that the plan was obviously designed to destroy the nonminority demographic establishment in Cook County. This argument was overruled and the courts accepted the more than proportional plan. A similar story unfolded in California where a plan had been prepared that went very far toward maximizing the number of minority districts (see Figure 6.5). This plan, however, did not go far enough for the minority group leaders and thus was disputed in the courts. Although it is important to recognize the significance of long-term discrimination, it should be noted that we are risking a fundamental shift from the historical territorial basis of representation toward a pluralist idea of representation. The necessity to gerrymander for racial purposes effectively undermines arguments to avoid gerrymandering for political or other reasons. The only escape from racial gerrymandering is residential integration. But the racial basis for representation and power places a premium on maintaining racial segregation.

The issue of race-conscious districting remains in the forefront. The ill-fated nomination of Lani Guinier for a post in the attorney general's office revealed the ambivalence felt even within the racial minority communities regarding a strategy of racial exclusiveness; Guinier argued for fewer gerrymandered multiperson districts with transferable voting.[4] In June 1993, the Supreme Court ruled, in *Shaw* v. *Reno,* that the North Carolina congressional districting, which resulted in the election of two blacks to Congress, was, in effect, a deliberate segregation by race (see Figure 6.6). In several subsequent cases, courts have ruled against racial gerrymandering, that is, against districting for the overt purpose of maximizing minority representation, including cases in Georgia and Texas, and have forced the revision of districting plans. Despite the loss of majority black districts, black incumbents were reelected in the 1996 election, undermining the claim of the necessity of racial affirmative gerrymandering to enhance minority interests. Guinier's (1995) proposed solution of shifting to cumulative voting in multimember districts, a limited form of proportional representation, has received some favorable attention but has so far only been implemented at the local level—school districts, city councils, and county commissions. Cumulative voting permits voters to concentrate their votes, if desired, on one candidate, say, in a three-person district, thereby enhancing the chances for representation, not only for racial minorities but for other interests. So far, this has been deemed too radical at the congressional level, partly because the districts would be so large—1.7 million people.

Conclusion

Electoral districts are ephemeral. Increasingly, American citizens will be unable to describe their legislative or congressional district. Yet electoral districts profoundly affect the distribution of political power. To the extent that they do not *fairly* represent territories, political parties, or racial and ethnic communities, the structure of districts can help some and hurt others. Although malapportionment is no longer a problem, gerrymandering remains a prominent tool for discrimination, most commonly by one political party against the other or, in some areas, to discriminate against urban

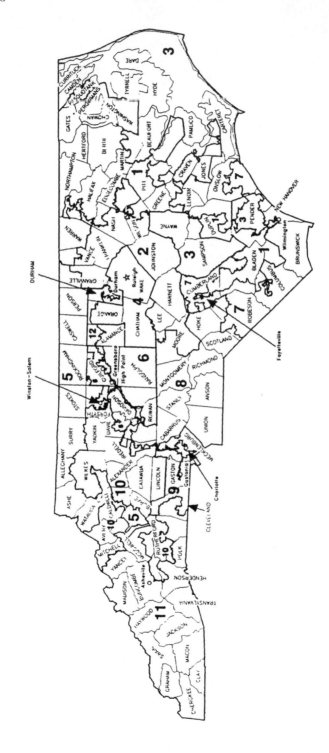

FIGURE 6.6 *North Carolina congressional districts, 1990. The North Carolina districts reflect the age of affirmative gerrymandering and led to the case of Shaw v. Reno.*

SOURCE: *Congressional Districts, 1990* (Washington, D.C.: Congressional Quarterly, Inc., 1990).

or rural territory or more liberal or conservative populations. On the other hand, the requirement for racial fairness has led to the use of affirmative gerrymandering to maximize minority group representation. All this has had the effect of diluting the historic meaningfulness of electoral districts, as communities with identifiable interests. Recent court decisions suggest a reaction to the loss of a sense of geographic integrity, as more mention is directed to community of interest and the role of political units such as cities and counties. But it is fundamentally difficult to reconcile competing goals of maintaining a meaningful form of territorial representation and electoral fairness.

Notes

1. Robert Dixon, *Democratic Representation: Reapportionment in Law and Politics* (New York: Oxford, 1968).
2. Richard Morrill, "A Geographer's Perspective," in Bernard Grofman (ed.), *Political Gerrymandering and the Courts* (New York: Agathon Press, 1990).
3. Richard Morrill, "Making Redistricting Models More Flexible and Realistic," *Operational Geographer* 9 (1991): 2–10.
4. Lani Guinier, *The Tyranny of the Majority* (New York: Free Press, 1995).

References

Baker, G. "Judicial Determination of Political Gerrymandering: A Totality of Circumstances Approach." *Journal of Law and Politics* 3 (1986): 1–19.
Cain, B. "Assessing the Partisan Effect of Redistricting." *American Political Science Review* 79 (1985): 320–333.
Garfinkel, R., and D. Nemhauser. "Political Districting by Implicit Enumeration Techniques." *Management Science* 16 (1970): B405–B508.
Goodchild, M., and B. Massam. "Some Least-Cost Models of Spatial Administrative Systems." *Geografiska Annaler* 52B (1969): 86–94.
Grofman, B. "Criteria for Redistricting: A Social Science Perspective." *UCLA Law Review* 3 (1986): 77–184.
Grofman, B., ed. *Political Gerrymandering and the Courts.* New York: Agathon, 1990.
Gudjin, G., and P. Taylor. *Seats, Votes, and the Spatial Organization of Elections.* London: Pion, 1978.
Hardy, L. *The Gerrymander: Origin, Concepts, and Reemergence.* Claremont, CA: Rose Institute of State and Local Government, 1990.
Hess, S. "Nonpartisan Political Redistricting by Computer." *Operations Research* 14 (1965): 998–1006.
Johnston, R. *Politics, Elections, and Spatial Systems.* New York: Oxford, 1979.
Morrill, Richard. "Political Redistricting and Geographic Theory." *Resource Publication in Geography.* Washington DC: American Association of Geographers, 1981.

_____. "Redistricting, Region, and Representation." *Political Geography Quarterly* 6 (1987): 241–260.

Niemi, R., and J. Deegan. "A Theory of Political Districting." *American Political Science Review* 72 (1978): 1304–1323.

O'Loughlin, J., and A. M. Taylor. "Choices in Redistricting and Electoral Outcomes: The Case of Mobile." *Political Geography Quarterly* 1 (1982): 317–340.

Schwartzberg, J. "Reapportionment, Gerrymandering, and the Notion of Compactness." *Minnesota Law Review* 50 (1966): 443–457.

Shelly, F. "A Constitutional Choice Approach to Electoral Boundary Delineation." *Political Geography Quarterly* 1 (1982): 341–350.

Part Two

Political Geographies of People and Resources

Chapter Seven

Population, Politics, and Geography: A Global Perspective

GEORGE J. DEMKO

Since the end of World War II, there has been a very rapid growth in awareness of, and concern for, the global "population problem." This chapter examines the population problem by disassembling it into its component
parts and assessing these parts from a geographical perspective. At the
same time, population growth and distribution issues are discussed in terms
of their impact on international relations. These relations involve, of
course, not only sovereign states but also intergovernmental (IGOs) and
nongovernmental organizations (NGOs).

The population problem all too frequently is viewed as an aggregate
global population issue—that is, a population menacingly large and growing at a rate that imperils the global environment, food supply, and resource base. The planet's population in 1997 was estimated to be more
than 5.84 billion, growing at a rate of 1.5 percent per year. On average,
every woman on earth will give birth to 3.0 children in her lifetime at current rates. All of this information has the ring of catastrophe about it and,
indeed, has stimulated an increasingly shrill debate among scholars and
policymakers. This debate cannot be evaluated, however, without first understanding that there are *a number of different population problems,* each
with its political ramifications.

The population problem has at least five significant components, each of
which is significant. These problems include (1) population growth and/or
lack of growth, (2) the population distribution problem, (3) the issue of in-

ternational population flows (migration of all kinds), (4) mortality and morbidity rates of population, and (5) cultural intolerance by governments and their populations (racism, gender discrimination, etc.). Clearly, these problems are interrelated and complex; they also vary greatly from place to place and have significant international implications.

The issue of global population growth was identified early by the English parson and economist Thomas Malthus. His famous prophecy of population growth outstripping the means of subsistence continues to influence some social scientists and is vehemently rejected by others. Among the most outspoken examples of these demographic extremes are Paul Ehrlich and Julian Simon. The former argues that population is indeed a socioeconomic "bomb" threatening the planet, whereas the latter argued that population growth stimulates economic growth. The population argument raises such related issues as maximum carrying capacity of the planet, the rate of global environmental degradation, the role and growth of technology and human ingenuity, and much more.

Clearly, the question of which argument is correct has no immediate, definitive answer. It would appear reasonable to advise the planet's decision-makers to err on the more conservative side and promote efforts to check population growth. But from a geographic perspective, it is more important to view the issue more specifically from a regional or disaggregated point of view. In short, there are indeed some places in the world where population growth is too high and is a serious problem; conversely, there are other places on the globe where a lack of population growth is a serious concern. The global aggregate view is distorting and must be clarified with a region-specific analysis.

Even at the global level, the rate of growth of population has slowed— the total fertility rate has dropped from 5.0 in 1950 to 3.0 in 1997.[1] The problem is that high fertility rates of the 1950s, 1960s, and 1970s are reflected in high population growth rates in the 1990s when the children born 20 to 30 years ago reach childbearing age. To overcome this "lag" or momentum effect, a fertility rate decline must be sustained for a long period to produce a decline in population growth.

Generally, there is a consensus that a "demographic transition" has been occurring over various parts of the globe as modernization, urbanization, secularization—prosperous development—change attitudes about family size, marriage, and so forth. Nuances of this model are emphasized by those who argue for the importance of the status of women and the education of girls as critical modernization variables. There has also been considerable discussion concerning the effectiveness of family planning in this historical process.

At a disaggregated level, however, there is remarkable variation in the "population growth" problem. The data in Table 7.1 present a sample of

TABLE 7.1   Total Fertility Rates for Selected Regions and Countries, 1997

	Total Fertility Rate
More developed countries	1.6
Less developed countries	3.4
Africa	5.6
Asia (excluding China)	3.5
Latin America	3.0
Europe	1.4
Niger	7.4
Malawi	6.7
Uganda	6.9
Iraq	5.7
Syria	4.7
Yemen	7.2
Philippines	4.1
India	3.5
Iran	4.7
Singapore	1.7
China	1.8
Japan	1.5
Mexico	3.1
Honduras	5.2
Cuba	1.5
Argentina	2.8
Italy	1.2
Germany	1.3
Russia	1.3
United States	2.0

SOURCE: *Population Data Sheet* (Washington, D.C.: Population Reference Bureau, 1997).

this variation. Obviously, a number of places are experiencing a severe problem of high growth. These states are predominantly in the Third World, and their problems of poverty and low quality of life are exacerbated by population growth. It is also clear that a number of states have made enormous progress in reducing rates of growth (e.g., China and Singapore), and much can be learned from these examples. Even within states, large variations are discernible and reflect spatial differentiation in the fertility reduction process. And for a growing number of states, low population rates compounded by aging populations are cause for concern. The population of Japan, for example, will begin to decline in 2011. Nearly sixty of the world's states have reached or dropped below replacement level growth rates, a situation with implications for international migration, labor issues, and other processes. Many of the developed states will be faced with aging populations, which will affect labor demand, pension funds, and may lead to a renewed round of labor immigration from the Third World. The latter may become a very sensitive issue, given the problems experienced in Western Europe already with guest workers and the growing xenophobia in a number of countries.

The variation in population growth rates described above may clarify the population growth problem but only provides a clear focus on those areas that require some type of "population aid." In fact, the specter of these populations growing at unacceptable rates led to a global effort to develop policies and programs to ameliorate the problem. By 1965, many developed states were providing population aid in the form of family planning support to interested developing countries. The U.S. Agency for International Development (USAID), for example, was given a mandate to provide technical assistance for population control to interested governments, primarily in the developing world.

In 1971, a World Population Plan of Action was created in which many developing states stressed the "value of people" to emphasize their right to decide on population issues. The Bucharest global population conference in 1974 reached a consensus in recognizing the value of family planning but, with the urging of the Soviet bloc in those Cold War years, insisted on the need for "socioeconomic transformation"—a euphemism for the need to invoke political reform of a Marxist type. By 1981, the direct external assistance for population programs to developing countries was more than $500 million a year, and the United States provided about 40 percent of this total. In 1984, at the Mexico City Conference on Population, the majority of the world's states insisted on the importance of population aid programs for developing countries funded by the West. The United States, however, injected a note of restraint regarding the use of abortion and coercion in family planning programs, and the population issue became remarkably politicized. The policy ostensibly was aimed at curbing coercive family planning programs and promoting a "humane system" of options for women. Many specialists and others around the world, however, argued that it was an attempt to impose prevailing American morality on others and resulted in depriving women of their rights.

The impact of the U.S. policy was great. Funds were cut off to a number of very important intergovernmental and nongovernmental organizations such as the United Nations Fund for Population Activities (UNFPA) and International Planned Parenthood Federation (IPPF), and tensions arose between these groups and the United States. China was affected by the UNFPA cuts and the incident added to the foreign relations strain between China and the United States. A number of West European states expressed their disagreement with the U.S. policy, and the Soviet Union stooped to the occasion by making its first-ever donation to UNFPA—in hard currency! This demographic issue remains highly politicized; although the U.S. government formally rescinded its Mexico City Population Policy in 1993,[2] the funding issue is now tied to the repayment of U.S. arrears to the United Nations.

The current period is marked by a shift toward a concern for population growth and the environment. Threats to the tropical rain forest, biodiversity, and related environmental issues have focused attention on developing countries and their rapidly growing populations in need of land and resources. On a per capita basis, greater environmental damage is inflicted by high levels of consumption in developed states than by actions of populations and governments in the Third World. It has been politically sensitive to bring pressure to bear on Third World states over their high population growth rates. This concern was finessed weakly in the Rio Conference on Environment and Development in Principle 8 of the Draft of Environmental Rules: "To achieve sustainable development and a higher quality of life for all people, states should reduce and eliminate unsustainable patterns of production and consumption and promote appropriate demographic policies."[3]

The World Population Conference in Cairo in 1994 addressed population issues in a more realistic way and reached some consensus over measures needed to address the problems, including the source of funding for amelioration and solution. There is little doubt that population policies, global and national, will increasingly and overtly be directed toward development and environmental concerns.

Population Distribution

The United Nations, at the end of the 1970s, began a regular process of surveying its member states on their various population problems. Many were surprised to find that population distribution problems were identified as significant for an overwhelming number of countries, and for many it was noted as the most serious problem. In most cases, the problem was defined as a rural versus urban maldistribution of population, with the growth of large urban centers (primate cities) ranking first among the issues. Rapid urbanization, though, is often tied to rapid population growth in particular rural regions. Such maldistribution problems are always serious national concerns and increasingly impinge on the global stage. Political instability and insurgency and the use and diffusion of drugs spill out of such demographic cauldrons. The global urbanization process stands out as a major concern. Urban population growth in the Third World exceeds 3.5 percent per year—outstripping the overall population growth rate. Megacities, tied into the global economic system, are being created, which are difficult to manage, filled with poverty and crime, and plagued by infrastructure and housing shortages. Of the ten largest cities in the world, eight are in the developing world; and of the thirty largest cities, twenty-two are in the Third World. By 1990, there were 215 million households in cities without safe water and 340 million without adequate sanitation. During the 1970s, Third World cities had to house 30 million additional people each year, and by the 1990s, 60 million more each year.

Drugs, crime, pollution, insurgency, and violent ideologies are largely concentrated in today's megacities. Governments are demanding new and more research and policy aid from intergovernmental organizations such as the United Nations and other international bodies on how to manage congested, impoverished urban populations.

International Population Flows

Cross-boundary flows of population have recently reached magnitudes not experienced since World War II (see also Chapter 8). These flows are varied and include refugees and asylum seekers, immigrants (legal and illegal), contract laborers (guest workers), and even international circulators (individuals who cross international boundaries for many reasons with intentions of returning to their home countries). Currently, at least 100 million people around the world live in countries other than those in which they were born.[4] The bulk of this massive flow is from the South (developing world) to other parts of the Third World and, to a lesser degree, to the North (developed states). Flows of poor migrants and refugees from developing countries to other Third World states have exacerbated the problems of the developing world. The flows to developed states, where they can more reasonably be accommodated, has been significant and has caused much stress. From 1980 to 1985, the three traditional countries of immigration—the United States, Australia, and Canada—took in nearly 4 million persons. From 1985 to 1989, these three accepted 750,000 persons from developing countries alone—more than 70 percent of their total admissions. In 1992 alone, the United States accepted 810,635 legal immigrants. There are more than 13 million refugees in the world today, down somewhat from the 18 million of the early 1990s, and more than 25 million displaced persons still within their countries (more than 7 million in Africa). Events in the early 1990s produced tens of thousands of emigrants from the former Soviet bloc and more than a million asylum seekers to Western European states (mainly Germany) from the former Yugoslavia.

These massive spatial flows of people are becoming increasingly politicized, and many governments are altering their policies with regard to such flows. A former head of the U.S. Immigration and Naturalization Service noted that the increase in pressure from migrants "challenges the capacity of governments to uphold basic sovereignty, in this case, the choice of who resides in one's own country."[5] Mrs. Sadako Ogata, the UN High Commissioner for Refugees, stated that "migration must be treated not only as a matter for humanitarian agencies of the UN but also as a political problem which must be placed in the mainstream of the international agenda as a potential threat to international peace and security."[6]

Many intergovernmental and nongovernmental agencies have become deeply involved in these population flows. Agencies from the United Nations High Commission for Refugees to the International Rescue Committee have challenged governments and bankruptcy to provide aid and attempt to bring down boundary barriers. Governments have reacted variously, with most West European states raising barriers and tightening asylum and refugee rules. Canada, Australia, and the United States initially relaxed admission requirements but are in the throes of reevaluating asylum and immigration policies, as flows—legal and illegal—affect domestic economies. The near future would appear to hold little relief from such population flows. Increased economic development is bound to stimulate migration from developing states, and nationalism, which has generated much of the recent turmoil and population flight, is hardly on the wane. What is clear is the need for a greater effort on the part of all actors— IGOs, NGOs, and states—to coordinate efforts to ameliorate the pain of some flows and perhaps even find solutions for others.

Mortality and Morbidity

Rates of dying (or, alternatively, longevity) and levels of health (or, alternatively, morbidity) vary remarkably within and among countries. The variation is readily apparent in Table 7.2 where the greatest longevity gap is more than thirty-five years; similarly, infant mortality rates vary from a high of 136 to a low of 4. Although mortality and morbidity are normally considered the exclusive concerns of sovereign governments, these issues have also spilled over boundaries and have become global and politicized. Conditions of overpopulation and economic deprivation in a number of regions of the globe have become so persistent (Bangladesh, Somalia) or have been so exacerbated by despotic governments on segments of the population (Sudan, Iraq) that international efforts have been mounted to bring a measure of relief or to avert massive loss of life.

Although health has improved in most places on the globe over the past forty years, there is great unevenness in these gains. Sub-Saharan Africa, for example, has experienced relatively little improvement. Deaths from AIDS, malaria, and tobacco-related causes result in 2 million deaths a year. Aid for health has dropped from 7 percent of the total official foreign assistance in the 1980s to 6 percent in the 1990s.[7] Hunger and related nutritional diseases affect millions of people around the globe. It is estimated that more than 500 million people, or 10 percent of the world's population, are undernourished. Contrary to conventional wisdom, most of the malnourished

TABLE 7.2   Life Expectancy at Birth and Infant Mortality for Selected Regions and Countries, 1997

	Life Expectancy at Birth	Infant Mortality (Infant deaths before age 1 per 1,000 live births)
Developed countries	75	9
Less developed countries	63	64
Africa	53	89
Asia (excluding China)	63	67
Latin America	69	39
Europe	73	10
Guinea	45	136
Kenya	54	62
Uganda	41	81
Israel	77	7.2
Yemen	59	79
Afghanistan	43	163
Sri Lanka	72	17.2
Singapore	76	4.0
China	70	31
Japan	80	4
Costa Rica	76	13.3
Mexico	72	34
Haiti	50	48
Sweden	79	4.2
Russia	65	21.2
Turkmenistan	66	46
United States	76	7.3

SOURCE: *Population Data Sheet* (Washington, D.C.: Population Reference Bureau, 1997).

suffer from "silent hunger," or chronic malnourishment, as opposed to famine. Of the eleven major famines since 1960, seven were caused primarily by civil war, interstate war, or government policies, not by natural catastrophe or overpopulation. Silent or chronic hunger, which affects more than 95 percent of the hungry, does not necessarily kill.[8] It raises morbidity by increasing susceptibility to diseases, raising infant mortality, and maiming victims physically and mentally. Chronic hunger is closely related to poverty and has a myriad of causes. It is found in places of poor resource endowment, high population growth, severe resource inequality, low and/or inappropriate agricultural technology, and combinations of all these factors.

Spatially, hunger exists everywhere, but its greatest concentration is, of course, in the Third World. The "hunger belt," focused on sub-Saharan

Africa and South Asia (Bangladesh, Bhutan, India, Nepal, Pakistan, and Sri Lanka) is the most critical set of regions of need. There are, however, other pockets of hunger in East Asia and the Pacific (40 million people), Latin America and the Caribbean (50 million), and the Middle East and North Africa (10 million). Much of the difficulty in resolving the hunger issue is not in a global lack of food but rather in a myriad of barriers to distribution, including economic and political barriers at subnational and international levels. A study by the International Institute for Applied Systems Analysis (IIASA) estimates that if all the hungry were located in one place, it would take about \$21 billion to feed them adequately, or only \$20 per person.[9]

Although there is enough food in the world to feed all the hungry adequately and there is an extensive international aid system to feed the hungry, hunger persists. The solution, of course, lies in providing the hungry with the means to feed themselves; but achieving such a noble goal requires complex and place-specific sets of programs. In many cases, it requires a change in government, infusion of technology, massive infrastructure development, and myriad measures necessary to promote economic development in the poor regions of the world. It may also require lower population growth rates in some areas, so that improvements in agricultural production are not offset by high birth rates.

A large number of international governmental organizations (e.g., FAO, World Bank, WHO), a myriad of nongovernmental organizations (Bread for the World, World Hunger Program), and many national governmental programs work diligently to address problems of unnecessary mortality and excessive morbidity. Many operate in conflict with states, and often they have only scarce resources to bring succor to the needy. All too often these well-meaning efforts fall short because of lack of cooperation and coordination with local agencies and governments.

AIDS

One of the most ominous plagues in modern times is AIDS and its precursor, HIV infection. The disease raises victims' morbidity level and is, at this point in time, always fatal. By 1997, 30 million people worldwide were infected with the virus, and the projections for the future are horrendous. Fully 90 percent of the infected population is located in developing countries.

The disease had, and still has, enormous political, geographic, and economic implications. During the Cold War, Soviet propaganda accused the United States of developing and spreading the virus and maliciously dispersed this misinformation in badly afflicted sub-Saharan Africa. A number of foreign governments in the 1980s, seriously frightened by the high rate of incidence in the United States, began to demand inspection rights over U.S. military personnel at U.S. bases. A number of countries rapidly altered their immigration rules, demanding special tests and rules for HIV- and AIDS-infected persons. More recent concerns focus on the continued rapid rate of growth of cases and especially the rate and spatial diffusion in certain parts of the world. The incidence of HIV and AIDS among males, especially homosexual men and intravenous drug users, has increased mortality rates of males aged 18–35 in the United States. Clearly, the impact of AIDS and HIV differs greatly from group to group and from region to region, demanding that solutions and even policies be population- and place-specific. In sub-Saharan Africa, the heterosexual diffusion of the disease has reached enormous levels and threatens large segments of the populations in Uganda, Zambia, and other countries. The estimated number of deaths in Africa from AIDS is 500,000. In South and Southeast Asia, the incidence of AIDS and HIV has increased remarkably, threatening untold numbers of people in such places as Thailand and India.

Given that types and methods of diffusion of the disease vary widely from place to place and that there has been a rapid increase in the variety of genotypes, a solution to the problem at any time soon is unlikely. The need for an unprecedented and massive international effort is obvious but, as yet, unfulfilled. The totality of issues affecting global and regional rates of morbidity and mortality from HIV and AIDS are serious and connect the populations of every place on the globe. As we approach the end of the twentieth century, state governments, intergovernmental organizations, and nongovernmental organizations have a moral obligation to address hunger and public health threats such as AIDS and the other deadly viral and bacterial afflictions in a more efficient, coordinated, and cooperative manner.

Cultural Intolerance by Populations

Most demographic textbooks include a section on demographic characteristics of population. The "acquired" characteristics such as religion and education are discussed along with "biological" characteristics, such as age, gender, and race, and they are related to the fundamental demographic processes—fertility, mortality, and mobility. These characteristics may be even more important in terms of how certain populations treat or mistreat each other. At this point in time, there has occurred an enormous explosion in, and awareness of, the many types of intolerance that afflict our planet. The problems vary widely from country to country and region to region—from gender discrimination in nearly every part of the world to racial, religious, and ethnic hatred and intolerance of many groups, in most countries of the world. These problems have been intensified and magnified by a number of global geographic and political processes. The enormous flows

of international migrants over international boundaries—from legal and illegal movers to refugees, tourists, and others—have greatly increased the
contact between dissimilar groups. The millions who have fled political repression are being replaced or supplemented with hundreds of thousands
fleeing the upheaval caused by the collapse of former Marxist states. Ancient religious and ethnic hatreds, once controlled by coercive governments,
have flared into bloody confrontations over territory in the former Yugoslavia and U.S.S.R. Many flee the violence and economic insecurity to
countries where they are confronted with hostile populations concerned for
their own welfare.

Examples of cultural intolerance are varied and range from xenophobic
hostility to immigrant populations in Western Europe in the early 1990s
(especially Germany), ethnic civil war among Croatians, Serbs, and Muslims in their former Yugoslavia, to the aborting and infanticide of female
fetuses and babies in South and East Asia. Gender discrimination reached
such a level of awareness in the early 1990s that one of the principles of the
Environmental Rules from the Conference on Environment and Development in Rio in 1993 called for the "full participation (of women) in environmental management and development." In 1993, the Canadian government extended "asylum" to a Middle Eastern woman whose claim of
gender persecution by her own government was found valid, setting a new
and important precedent in asylum practices.

The first international conference on human rights, sponsored by the
United Nations and held in Vienna in June 1993, was remarkable for its tu-
multuousness. Thousands of oppressed groups and human rights NGOs
gathered outside the main meeting halls to display evidence of cultural intolerance and resultant persecution in every region of the world. The conference demonstrated clearly the significance of these issues and the relative
impotence of the UN in taking effective measures to alleviate the problems.

Conclusion

Although population problems are most often debated at the national level,
like environmental processes, they are no longer confined within the
boundaries of states. Population pressures indeed respect no boundaries
and impinge on the global community, engaging all of the global actors.
"Overpopulation" in one region of the world impacts all the world's states
in the form of resultant political instability and outflows of migrants—from
illegal "economic" migrants to asylum seekers. Such movers impinge on
states with aging populations and high levels of economic development and
increasingly result in exclusive border controls meant to reject foreigners.
Burgeoning megacities affect the "global village" in a similar way, in that
they are intimately connected to all the world's population via the spatial

diffusion of drugs, crime, AIDS, terrorism, and more. Cultural intolerance, whether in terms of gender, race, religion, ethnicity, class, or any other form, produces only more hatred or civil turmoil.

Many of the demographic issues discussed above are or can be interrelated. For example, lowering birth rates can be tied to better public health programs in family planning centers in rural areas. Similarly, such programs can be bases for educating women and girls, a strategy known to lower fertility and improve health levels. Finally, such programs and centers can serve to distribute condoms and to educate populations about AIDS and HIV.

As a New World Order of some type emerges at the onset of the twenty-first century, it will certainly be influenced by how the world manages its population. International instruments and governance will rapidly increase in number and significance. It would seem obvious that in such a spatially interconnected world, states, IGOs, and NGOs must address demographic issues more collectively. Cooperation and coordination are mandatory, if only to ameliorate the high price of political-demographic problems. These issues are too important to allow the current chaos and unilateral decision-making on issues that determine the death and life of millions of humans. Clearly, such international efforts can address concerns of cost, definitions, aid, policy coordination, and more in a manner worthy of a twenty-first century global community. In a world of nearly 9 billion people foreseen by the year 2025, there will be even greater pressure on our resources (especially water), probably less tolerance, and less latitude for political decisions. There has never been a more propitious or critical time for coordinated planning based on an accurate geographic perspective on population problems.

Notes

1. The total fertility rate (TFR) is the number of children a woman will have, on average, in her lifetime (measured at current birth rates). The level of population replacement is represented in a TFR of approximately 2.1.

2. "Recession of the U.S. 'Mexico City Policy,'" *Population and Development Review* 19 (March 1993): 215–216.

3. "Draft of Environmental Rules: 'Global Partnership,'" *New York Times,* April 4, 1992, p. A6.

4. "Migration Report," United Nations Population Fund, July 6, 1993.

5. Doris Meissner, "Managing Migrations," *Foreign Policy* 86 (Spring 1992): 68.

6. G. Lyons and M. Mastanduno, *Beyond Westphalia? International Intervention, State Sovereignty, and the Future of International Society* (Hanover, N.H.: Dartmouth College, 1992), p. 20.

7. *World Development Report, 1993* (Washington, D.C.: World Bank, 1993).

8. G. Fisher, K. Frohberg, M. A. Keyzer, K. S. Parikh, and W. Tims, *Hunger: Beyond the Reach of the Invisible Hand* (Laxenburg, Australia: International Institute for Applied Systems Analysis, October 1991).

9. Ibid., p. 37.

References

Demko, G. J., and R. J. Fuchs. "Population Redistribution: Problems and Policies." *Populi* (United Nations) 7, no. 4 (1981): 26–35.

Ehrlich, P., and A. Ehrlich. *The Population Explosion.* New York: Simon and Schuster, 1990.

Gobalet, Jeanne G. *World Mortality Trends Since 1870.* New York: Garland Publishers, 1989.

Keyfitz, N. "Population and Development Within the Ecosphere: One View of the Literature." *Population Index* 57, no. 1 (1991): 5–22.

Malthus, Thomas R. *First Essay on Population, 1798* (A Reprint in Facsimile). London: Macmillan, 1966.

National Research Council, Committee on Population, Working Group on Population Growth and Economic Development. *Population Growth and Economic Development Policy Questions.* Washington, D.C.: National Academy Press, 1986.

OECD/SOPEMI. *Trends in International Migration: Continuous Reporting System on Migration.* Paris: OECD, 1922, 1995, 1997.

Simon, Julian. "A Scheme to Promote World Economic Development with Migration." In *Research in Population Economics,* ed. J. Simon and R. Lindert. Greenwich, Conn.: JAI Press, 1982.

_____. *The Ultimate Resource.* Princeton, N.J.: Princeton University Press, 1981.

United Nations Fund for Population Activities. *State of the World's Population.* New York: UNFPA, 1992.

United Nations Fund for Population Activities. *Population and the Environment: The Challenges Ahead.* New York: UNFPA, 1991.

World Health Organization. *Our Planet, Our Health: Report of the WHO Commission on Health and Environment.* Geneva: WHO, 1992.

Chapter Eight

International Migration: One Step Forward, Two Steps Back

WILLIAM B. WOOD

Governments face the new millennium with few good answers on how to best tackle an age-old dilemma, the uncontrolled movement of people into and out of their sovereign territories.* Although Western governments and politicians have been the most vocal about the woes of illegal immigration and the high cost of supporting refugees, in reality poor countries are more profoundly affected by international migration flows of all kinds. Poor countries face the additional and much larger problem of coping with the impacts of large migrations within their borders, in particular that from rural to urban areas. For governments of these countries, international migration is the tip of a population redistribution iceberg, which will have profound political stability and economic development consequences for the next century.

A discussion of the political geography of international migration requires first a quick review of migration theory because of similarities between subnational and international migration flows. All migrants share many of the same social, cultural, political, ecological, and economic forces that lead individuals and groups to leave one area for another. International migrants are distinguished from subnational migrants less by their motives

---

*The views in this article are those of the author and not necessarily those of the U.S. Government.

than by their crossing of an often arbitrarily drawn international boundary and their coping with a different set of legal, economic, political, and cultural systems than those they experienced in their home country.

Historically, international migration—whether forced or voluntary—has served as a major conduit among countries, often permanently changing their population compositions, economies, and even landscapes. In an increasingly interdependent world economy, which is also experiencing resurgent nationalism and entrenched poverty, international migration poses a difficult challenge to both individuals and governments. A migration stream from one country to another, for example, might be viewed from two quite different perspectives: for the migrants themselves, it can be a once-in-a-lifetime opportunity to escape bleak or even life-threatening conditions in their home country; for some politicians in the receiving country, these same migrants can represent a host of social and economic ills, from unemployment to crime. As debates over immigration and refugee policies intensify in national and international fora, a political geographic perspective can shed light on the dimensions and trends of international migration flows, important causal factors, and likely repercussions.

Migration Theories

Two of the three basic demographic variables, births and deaths, are relatively easy to monitor because they each occur only once per lifetime and at one place and time. But the third, migration, has proven much more frustrating to demographers because it can be experienced many times during a lifetime, among many different places, and for varying durations. Individual moves, for example, are usually only counted in a decennial census if a change of residence is involved and if a political boundary (municipal, county, state, or national) is crossed, leaving a sporadic and incomplete migration record. Perhaps because of its inherent complexity, migration has defied attempts at a generalized theory.

Mobility involves many forms and patterns and hence can yield different ways of classifying migration types. None are completely satisfactory because of the broad range of movements people engage in: commuting to and from work or school, a daily practice almost all of us undertake; nomadic herding and shifting cultivation, ancient practices that have allowed small groups to survive harsh environmental conditions; seasonal migration, involving crop planting and harvesting cycles and off-farm employment; and rural to urban migration, a shift in population distribution that continues to change the socioeconomic foundation of every country. These diverse mobility patterns result from a range of causes, involve many different actors and institutions, and result in varied repercussions on the places of origin and destination.

Since E. G. Ravenstein set down his "laws of migration" more than a century ago, social scientists have pondered the linkages between migration flows and a laundry list of influencing variables: age, gender, distance, economic conditions, and rural/urban wage disparities. Migration can be usefully described as a process determined by "push and pull" variables that operate at both the origin and destination places.[1] In between these two places, "intervening obstacles" affect the ability of a migrant to complete his or her journey. Not all obstacles are physical, nor is the distance between places the primary deterrent in undertaking a journey, as it probably was in Ravenstein's time. Anyone who has bought an airline ticket can testify to the weak correlation between distance and travel cost. Today, the major obstacle to international migration is more likely to be governmental red tape than vast distances.

From a historical and geographic view, migration can be viewed within a broad "mobility transition," in which evolving patterns of mobility are tied to stages of economic development. Under this model, "advanced" and "super-advanced" societies require much less mobility than "transitional" societies, largely because of advances in telecommunications.[2] However, despite "information highway" technologies that have sharply increased transnational telecommunications—and hence should decrease work-related mobility in "advanced" societies—many people in the world, particularly in poor regions, continue to move in order to survive.

Over the past two decades migration research has focused on the selectivity of migration: Why do some people migrate while others stay put? This line of research has involved surveys of migrant decisionmaking, kinship linkages, communication networks, and perceptions of the pros and cons of moving to a new place. Migration theorists have looked closely at rural-to-urban migration, which has involved the radical transformation of rural peasants into city dwellers. Growing numbers of urban residents reside in crowded and sprawling squatter settlements, resigned to accepting low wages and intermittent work in an unregulated "informal" sector. In the early 1970s, the debate on rural-to-urban migration was sharpened with the use of migration decisionmaking models. The best known of these economic models was based on estimates of expected future earnings from an urban job, weighed against current real income and the estimated cost of moving. Many empirical studies have since underscored the importance of perceived economic opportunities in migration decisionmaking, whether the intended journey is across town or across an ocean.[3]

Sociologists and political scientists, however, find these economic models to be too deterministic; the models do not adequately address the role of formal institutions, informal networks, and government policies in either facilitating migration streams or diverting them. They argue that migration must be viewed within political, social, and economic contexts, especially when looking at international migration.[4] International labor flows from "peripheral" to "core" countries within the global economy, for example, respond to both economic incentives as well as the fluctuating interests of governments. The governments of industrialized countries might sponsor labor recruitment campaigns during a period of economic growth and then a few years later begin expelling unwanted workers when domestic unemployment rises (such as Germany with Turkish "guest workers" or the United States with Mexican "bracero" workers). Immigration policies can also be driven by noneconomic factors, such as ethnic relations. The Chinese Exclusion Act of 1882, the U.S. Quota Law of 1921, and the National Origins Act of 1924 are notable examples of how an explicit ethnic-based national policy can stop or divert international migration flows, in this case effectively halting the entry of non-European immigrants into the United States for several decades.[5]

Population geographers studying migration have focused on migration patterns, especially between rural and urban places and as part of a diverse set of economic development conditions.[6] They have looked at how migration affects the hierarchical development of human settlements and how various types of migration flows—from seasonal (usually based on fluctuating demands for farm labor during the year, with a peak at harvest time) to circular (with migrants moving back and forth between urban and rural residences depending on income-earning prospects)—are economic survival strategies in poor countries. Migration and urbanization are closely intertwined spatial processes, particularly in many poor countries where large rural populations provide a steady supply of migrants to rapidly growing cities. Fieldwork has shown how low-income migrants shuttling between tiny farms, villages on the urban periphery, and inner-city squatter settlements blur the distinction between urban and rural economies. Geographers have demonstrated how government policies directed at economic development goals have also had unanticipated, and often undesired, impacts on these migration flows.[7] Finally, geographers have looked at a range of factors and policies affecting refugees and other forcibly uprooted peoples, including the deliberate destruction of their environments.[8]

The study of migration will remain difficult because the departure of migrants from one area and their arrival in another generates fundamental changes to local labor markets, resource use (especially land), and economic production. Migrants bring with them their own cultural traditions, social structures, and ideologies that can define how they adapt to their new communities. The transient status of migrants also usually makes them marginalized political actors and thus more vulnerable to onerous government policies, economic exploitation, and discrimination by resident groups. All these interactions underscore why migration requires multidisciplinary analysis.

International Migration

International migration entails crossing an international boundary for an extended period of time (i.e., tourism is not usually included). As in other types of migration, the time frame is deliberately ambiguous because a migration journey might be several months or permanent. The migrants themselves may have little information on possible destinations, only a vague plan of how to get there, and little idea how long they will stay once they arrive. Governments increasingly view these migrations as national security problems and have made immigration the subject of high-level foreign policy debate.[9]

The UN estimates that more than 100 million people could be classified today as international migrants because they are resident in a country other than their place of birth. Although an impressive number, it needs to be viewed in context; with a global population of 5.8 billion, international migrants represent less than 2 percent of the world's population. Moreover, this estimate is only a ballpark figure because measurement of cross-border movements is haphazard at best. Border guards and immigration officials may record a particular crossing for someone who is granted official permission to enter. But once the border is crossed, most migrations within and perhaps out of the new country will likely go unrecorded. Illegal migrations are even less well understood or measured. Statistics on interdictions and deportations of illegal "aliens" reflect only the minority that are caught and depend on the level of government efforts to catch them.

Despite the lack of accurate immigration statistics and the fact they actually make up a relatively small percentage of most national populations, international migrants garner disproportionate political attention. Immigration is an especially sensitive political topic in Europe, in part because many Europeans do not view themselves as living in historically "immigration" countries. Efforts to form a more integrated and "borderless" European Union (EU), for example, may be hampered by fears that Third World asylum seekers will take advantage of intra-EU border controls weakened by the 1995 Schengen Agreement. Nationalistic political parties in Europe have used these fears to advocate much stricter immigration and asylum laws and to postpone lifting of restrictions on labor mobility within the EU.[10]

Scholarly difficulty in generalizing about migration patterns and trends is reflected by inconsistent government policies, especially over immigration. Debates over the costs and benefits of immigration, as well as appropriate and fair measures to control the number of "illegal aliens" in the United States, for example, will continue well into the next century.[11] About 10 percent of the U.S. population is foreign-born, about 26 million residents, with 27 percent of them born in Mexico, followed by the Philippines and Hong Kong/China (about 4 percent each). The 1986 Immigration Reform and Control Act helped legalize the status of about 2.7 million "undocumented" immigrants, mostly from Mexico, but did not solve the problem of illegal immigration. Thus, the U.S. Border Patrol along the Mexican border has been significantly bolstered in the 1990s to apprehend and promptly return those attempting to enter illegally. Despite these legislative and law enforcement efforts, several million "illegals" remain employed in the United States, generating heated political debates in several states over their access to basic social services, such as schools and health care. The national immigration debate also profoundly affects the prospects of a diverse group of Latino communities who share different migration experiences yet often are stereotyped by the Anglo majority.[12]

Within both the United States and Europe, distribution of immigrants varies widely; in the former, most are concentrated in six states (California, Florida, Texas, New York, Illinois, and New Jersey), and in the latter, the percentage of foreign born ranges from about 1 percent of the total population in Italy to almost 19 percent in Switzerland. Although the largest outcries over illegal immigration come from these industrialized countries, the largest numbers of undocumented migrants remain in poor agrarian countries. Similarly, the international migrants most often in the news are refugees; they, however, are increasingly outnumbered by those who have also been forcibly uprooted from their homes but have been unable to leave their country. Regardless of where they were generated, refugees and other forcibly displaced persons represent societies torn apart by civil war and widespread human rights abuses. Although political and economic factors spurring international migration overlap, governments have often responded to them as separate processes.[13]

Economic Migrants

All migrants, including refugees, move within the context of an increasingly integrated and interdependent world economic system, but they are also swayed by cultural ties, social institutions, and governmental policies. Despite the powerful worldwide economic forces that push and pull international migrants, the impacts of their moves are felt primarily at the local community level. Cross-border economic migrants defy stereotypes: nomads, laborers, merchants, and business executives; all seek to improve their prospects by moving, temporarily or permanently, to another country. Their main problem is usually whether the country they wish to migrate to will let them in, and if they are allowed in, how long that country will allow them to stay. Immigration policies that will change their lives—as well as the demographic composition of the countries involved—may be based not on their economic needs or even potential contribution to the recipient country but rather on ambiguous and often unstated political agendas.[14]

In a purely economic world, there would be no political boundaries interfering with international labor migration. Only in the past century have
political boundaries become such a commonplace barrier to migration
streams. Until just the last few decades, most migrants moved through
frontier regions, which were often dangerous zones for travelers because
they were on the periphery of control by adjacent kingdoms. The clearest
physical boundaries were the high walls around castles; their drawbridges
served much the same purpose as today's immigration checkpoints at every
international airport: They regulated who could enter the "fortress" of sovereign territory. Indeed, the perceived benefits of migrants has fluctuated
greatly. In the mercantilist era, a large population was viewed as an economic and military asset, and thus sovereign controls were focused more
on emigration than on immigration. In the colonial era, migrants were
viewed as the means to extend imperial domain to the far corners of the
globe. In the era of industrialization, migrants were and are seen as a key to
lowering labor costs and thereby increasing profits. More recently, this
trend has been made even more complicated by the fact that while migrants
are seeking factory jobs in more developed countries, the factories themselves have been moving to less developed countries to take advantage of a
large and cheap labor pool.

Worldwide economic interdependence and, to a lesser extent, concerns
over human rights and environmental degradation are now challenging the
concept of unquestioned sovereign space (see Chapter 12). Capital is electronically transferred among world cities, with or without governmental involvement, and foreign investments are sought to help drive economic
growth. Many potential international migrants are merely trying to follow
these capital investments to where future incomes are the most promising.
Those potential immigrants who are willing to invest large sums in their
new homes, such as wealthy emigrés from Hong Kong, usually receive red-
carpet treatment. Those without wealth, such as East European Gypsies,
face a colder reception, particularly when jobs in the industrialized countries are scarce. Those who circumvent immigration laws are branded as
"illegals" who live under the constant threat of deportation. Those that
hire "illegals" themselves become implicated in an international net of illegal migration-related activities involving well-organized international
smuggling rings.

International migrants often play a vital economic role in the places of
origin and destination, as well as the places in between. Migrants are usually not the poorest or the richest members of their societies, and typically,
they have had more education than their stay-at-home peers. The emigra-
tions of educated young people have been portrayed as a "brain drain,"
which hinders economic development in their home countries. In labor-sur-
plus economies, however, even college graduates can find themselves unem-

ployed or underemployed at jobs well beneath their skill or educational level. These emigrants would probably argue that their talents are wasted if they stay put and that international migration is their only path to develop their individual potential. Those that do become immigrants in countries with relatively high wages, however, all too often find themselves working at low status jobs they would have shunned in their home countries.

The benefits of remittances from overseas migrants in stimulating local economic growth is also debated among social scientists, with some claiming that received funds are primarily used for nonproductive consumption. In lieu of other sources of income, however, remittances become a critical pillar of family, village, and even national economies, especially in such labor-export countries as Mexico, the Philippines, Bangladesh, and Egypt. When the hundreds of thousands of foreign workers who fled from Iraq and Kuwait in 1990 finally made it back to their home countries, they faced high unemployment rates and families who now had to get by without much-needed remittances.

Economic migrants are often treated as a large, almost bottomless, labor pool for national development plans. When there is a domestic labor shortage in an industrializing or oil-exporting country, migrants are allowed temporary residence as "guest workers." When the economy contracts during a recession and domestic unemployment rises, these guests become less welcome, regardless of how long they and their families have stayed (including children born there) and how much they have contributed to prior economic growth. Germany, which since the early 1960s has relied heavily on several million guest workers from the former Yugoslavia and Turkey, is now faced with finding jobs for young eastern Germans and is offering financial incentives for its southern European "guests" to leave. Amid vocal anti-immigrant sentiment, other Western governments are raising their requirements for legal immigration and toughening their deportation procedures for illegal immigrants.

Political Migrants

Refugees are international migrants expelled by persecution and/or civil war.[15] United Nations–directed efforts to protect and assist victims are greatly strained by large and fluctuating numbers of refugees worldwide, estimated to be about 14.5 million in 1997 (down from a high of 18 million in 1992). The United Nations High Commissioner for Refugees (UNHCR) now faces escalating costs and difficult logistics in providing immediate care and protection, possible third country resettlement, and ideally, safe repatriation (return to their homes) for all refugees. A major UNHCR concern continues to be the prevention of *refoulement,* the forced return of refugees. In addition, the UNHCR is beginning to cope with well over 20 million internally displaced persons (IDPs)—those fleeing persecution, war, or other life-threatening situations, but who remain within their country.[16]

Many of those forcibly displaced are members of oppressed ethnic groups deliberately removed from their homeland by state-controlled military forces. Some refugees, such as Pushtuns fleeing the 1979 Soviet invasion of Afghanistan, crossed an international boundary (into Pakistan) but remained within the traditional territory of their nation. Although large numbers of Afghans were repatriated, their homeland is far from stable, and renewed fighting could quickly create new mass flows of refugees and IDPs. Although the UNHCR in recent years has seen a leveling off of refugee numbers, it has seen a sharp increase in the numbers of IDPs who require UNHCR assistance but are without the benefit of UNHCR-mandated protections accorded to officially recognized refugees. Since the Balkan War in the early 1990s, Europe has joined Asia and Africa as a major refugee generating region.

The role of UN and nongovernmental refugee relief agencies has also become more entangled in political and military issues. The return of several hundred thousand Kurds to their homeland in northern Iraq in 1991 occurred relatively smoothly but required the presence of a defined safe haven area, a multilateral military force, and a UN guard force. International efforts to assist Iraqi Kurds, Somalis, Rwandans, and Bosnians raised difficult institutional problems for UN refugee and relief agencies, especially over claims of sovereignty infringement by member states on the one hand and on the other demands by human rights groups that the UN protect the principle of "humanitarian intervention" when oppressed groups, such as IDPs, are being massacred and regional stability is being undermined.[17]

Relief prospects for more than 3 million refugees and IDPs in the former Yugoslav republics improved following the 1995 Dayton Accords, but full repatriation as of 1998 remains frustrated. This dilemma has underscored the UNHCR's difficult role in providing relief and some measure of protection for those who have not crossed an international boundary but remain in a "refugee-like" condition. The brutal "ethnic cleansing" of many towns and cities in Bosnia pushed more than a million Bosnians into neighboring countries and into Western Europe (especially Germany), as well as corralling hundreds of thousands of IDPs into UN-protected "safe areas." Unlike refugees elsewhere, though, their expulsion was not a by-product of war but rather a deliberate military and genocidal strategy for controlling an ethnically homogeneous territory.

Chronic political instability and deeply embedded poverty will remain powerful but unpredictable generators of refugees in the Horn of Africa and Central Africa for the foreseeable future. In Somalia, the tragedy of civil war was compounded by a severe drought in 1991–1992, resulting in several hundred thousand deaths. Many of these Somali famine victims were nomads whose herds perished; more than 600,000 Somalis fled to refugee camps in Kenya and Ethiopia in the desperate hope of being fed. Famine and anarchy finally compelled the UN to authorize the use of a strong military presence to open up relief delivery corridors and feeding stations. In Rwanda, a refugee crisis was fueled by a vicious government-orchestrated genocidal campaign in 1994 against ethnic Tutsis. A successful Tutsi-led insurgency resulted in the expulsion of more than 2 million genocide perpetrators and ethnic Hutus into neighboring countries, where for more than two years they were fed by the international community. Although most Hutu refugees returned to Rwanda by 1997, their destabilizing presence in eastern Congo helped fuel first a regional rebellion and then a change of national governments in the distant capital of Kinshasa. As in other regions, refugee populations can fluctuate suddenly in both magnitude and location (see Figure 8.1).

Just as legal and illegal economic migration flows respond to perceived opportunities around the world, refugees fleeing political instability and oppression are barometers of change in geopolitical pressure. Although the cumulative number of economic and political migrants shows no sign of abating, as yet there is no international consensus as to where these migrants will go, who will protect their human rights, and how assistance will be provided to them.

International Migration Trends

Over the next few years, several migration trends will influence international relations: increasing mobility at local, regional, and international levels; growing complexity of causal factors that motivate migrants; and merging of internal and international migration patterns.

More Mobility

Despite increasing vigilance at international borders, the number of international migrants will likely remain high and may even continue to climb. If more rigorously protected borders around wealthy countries successfully dissuade or deter would-be illegal migrants, world mobility may become less visible but there nonetheless; instead of becoming part of the illegal international migration problem they will merely contribute growing numbers to the perfectly legal but perhaps even more unmanageable internal migration problem. The greater the magnitude of various migration streams, the more powerful their cumulative impact on social and economic institutions and on each other.

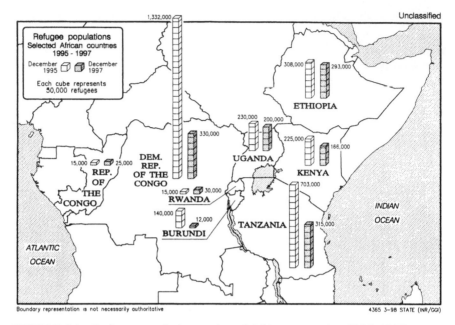

FIGURE 8.1   *Refugee populations, selected African countries, 1995–1997.*

SOURCE: Office of the Geographer, U.S. Department of State.

Multiple, Overlapping Motives

Although the motives that spur migrants to cross an international border are complex, immigration and refugee laws tend to look at international migration in black and white terms: A person who enters a country without permission and requests residence is either an illegal economic migrant or a political refugee.[18] Bureaucratically, this might make sense because it allows a simple criterion for refugee status: the presence of persecution or civil war. For some migrants, clear evidence might exist that would justify an asylum claim, such as when a military regime clearly targets a group for elimination. For many other asylum seekers, though, such clear evidence is lacking, which results in buck-passing among governmental agencies and tensions between the country where the migrant first applies for asylum and other potential resettlement countries.

The lack of simple and distinct causal factors in migration is forcing refugee and immigration agencies to look more closely at the relationships between places of origin and destination. Forced migrations can be caused by a broad range of factors: invasion by a foreign army or civil war; organized persecution of a minority group (usually by a government); and life-threatening hardship from a combination of economic and environmental threats. Recently, international agencies have begun to address the causes and consequences of "ecomigrations" that have been primarily influenced by adverse environmental conditions (natural and man-made). Often environmental, economic, ethnic, and political factors are closely intertwined, as they have been in the Horn of Africa, where militias have uprooted millions of peasants, killed off their herds, burned their crops, and stolen their food. Under such oppressive conditions, a drought can push hundreds of thousands of peasants and nomads over the edge from chronic malnutrition to death by starvation.

International and Subnational Linkages

The only significant difference between internal and international migrants is the obstacle of an international political boundary—an artificial construct that hinders the flow of people much more than it does the flow of capital, ideas, and goods. International migrations can also be viewed as an extension of several decades of unprecedented urbanization. Governments especially fear the rapid population shift from rural to urban areas. These migrants are literally urbanizing the world, most rapidly in poorer countries; over 43 percent of the world's population already live in urban areas. Crowded and impoverished urban populations can intimidate urban-based governments to keep food prices low, thereby weakening investments in rural agricultural development. Government policies that favor urban workers over rural farmers vary from country to country, but the net effect is to reinforce rural-to-urban migration trends helping to create ever more congested cities. These cities, particularly capitals, serve as critical stepping-stones for those migrating beyond a country's borders. "Global cities," with their closely interconnected business and financial networks, will likely remain the strongest magnets pulling international migrants.

The means that allow people to move—transportation networks, extended kinship connections, and loosely controlled job markets—are facilitating both internal and international migration. A world with much more mobility within countries will become one with much more mobility among countries, unless draconian immigration policies are ruthlessly enforced. Southeast Asia's financial woes in 1997–1998, for example, led to a harsh crackdown in Malaysia and Singapore against illegal immigrants, particularly those from Indonesia who have been rounded up, jailed, and deported by the thousands across the Straits of Malacca. Such efforts to discourage unwanted international migrations have had mixed results, with success at returning "illegals" usually only temporary.

Disorderly Flows

The next century's international migration and refugee flows might be better understood if they were placed within a reordered world. These new world categories are not based on ideological principles but on a wide disparity in economic opportunities among poor and rich countries. Within this world economic system framework, three global circuits of migration emerge: among poor countries; among rich countries; and between rich and poor countries.

Migration Among Poor Countries

Most migrants and refugees move among poor countries, which usually have per capita yearly incomes of under $600. Impoverished countries such as Malawi, Sudan, and Pakistan have hosted millions of refugees. Without substantial international support, these groups would pose an intolerable strain on local populations. Even with long-term international support, underlying socioeconomic conditions of chronic regional instability are rarely addressed. Some countries, such as Iraq and former Yugoslavia, which once had somewhat good prospects for economic growth, have been embroiled in wars, territorial nationalism, and racial/ethnic/tribal violence that have destroyed fundamental economic structures. Even if peace is restored and displaced groups are returned home, war-ravaged economies will take many years to recover and, given an opportunity, many of their young people will likely want to leave in the interim.

The poorest countries of the world will continue to bear a disproportionate share of the migration and refugee burden. A difficult problem for the UN and NGO relief agencies is how to assist them, without contributing to their permanent dependency on foreign assistance; many of the problems associated with returning economic migrants are dealt with by the Geneva-based International Organization for Migration. Deeply impoverished populations suffer directly from resource degradation and depletion (especially soil erosion and inadequate water supplies) and declining per capita agricultural productivity, trends that are very difficult to correct. If these widespread environmental scarcity concerns are not dealt with, mass displacement will continue; although most victims will remain within their country's borders, a sizable number will likely seek relief outside of them. Of particular concern are women, children, and the elderly, who are less mobile and thus less able to escape poverty and violence. Countless women and children will be victimized, girls forced into prostitution and young children into virtual servitude; without international human rights interventions, their plight will likely worsen.

Migration Among Rich Countries

Given worldwide economic problems, few countries are willing to admit that they are rich and are thus capable of keeping an open door for both unskilled migrants and refugees. Even the wealthiest of countries are highly selective of who can qualify for citizenship. Among the industrialized or "developed" countries, migration of highly skilled workers, managers employed by multinationals, and students help to more closely integrate these countries within an interdependent world economy. Migrants and travelers (including businesspeople and tourists) move among the global cities of the world, where identical hotels and fast-food restaurants help make temporary cultural transitions easier to endure. Excluding tourists, this is a small migration circuit—perhaps even in the tens of thousands—but it is important because it includes the world's key political and economic decision-makers.

Migration Between Rich and Poor Countries

The movement of people from poor to rich countries is perhaps the most contentious arena of migration policymaking. Like other international migrations, it, too, is complicated and may involve quite distinct rich, poor, and middle-class migration streams and counterstreams. Historically, the migration of Europeans to the far-flung colonies (or from the present-day "rich" to the present-day "poor" countries) established economic, political, and cultural linkages. Even within these historical migration streams, there were divisions between rich and poor, with wealthy migrants becoming major landholders and merchants in the colonies while the poor endured very harsh conditions as laborers. As with migration among rich countries, the migration of wealthy elites from poor or middle-income countries is usually not a problem because they bring money with them.

The migration of poor migrants from poor to rich countries, in contrast, is now viewed as a national security threat for many industrialized countries undergoing economic recessions because poor migrants bring with them only their labor and their aspirations. Much of the antiforeigner backlash in the new "Fortress Europe" can be traced to the perception that a "flood" of poor people (usually from former colonies) will take jobs away from locals. As unemployment rates climb, immigrants become easy scapegoats and immigration and asylum policies become highly charged local, national, and even geopolitical issues. Rebuffing previous policies that explicitly encouraged immigration of laborers, industrialized country governments have acted both unilaterally and multilaterally to restrict entry of impoverished job seekers and desperate asylum seekers, in effect building higher physical and bureaucratic walls between the world's haves and have-nots.

Global Disparity

These three generalized migration patterns reflect the one global trend that will drive most international migration flows to unprecedented levels: the growing economic and demographic disparity between rich and poor countries. Poor countries generally have low rates of economic growth and high rates of population growth. Few have reasonable prospects of creating enough domestic jobs to keep up with the basic demands of the current and next generation of workers. Frustrated young adults in these poor countries—some with college educations—understand all too well that these global disparities will influence their as well as their children's future. For many, the answer to this global inequity is simple in theory but increasingly difficult in practice: Move to a richer country with a smaller labor surplus.

Conclusion

Media attention on migrants has tended to emphasize the plight of a particularly desperate group of refugees or the latest proposal to curb the flow of illegal immigrants and has not adequately addressed the geopolitical and economic context in which migrations occur within and among world regions. The immigrations of Cubans into the United States in the 1960s and that of Indo-Chinese in the late 1970s and 1980s, for example, were the result of "national security" forays that have influenced at least two generations of immigrants as well as foreign policy relations with the generating countries. The political geography of these international migrations thus requires an understanding of how domestic and foreign policies influence migration patterns and trends. Regardless of where new migration flows occur, governmental efforts to deal with them will be complicated by economic disparities, regional political tensions, and mounting population and ecological pressures.

Many refugee policies in Europe and North America were conceived within a Cold War paradigm. Asylum was virtually guaranteed to anyone who successfully fled communism by crossing the "Iron Curtain" into the "Free World." With the Iron Curtain first falling and then removed, Western governments in the 1990s have been reevaluating their asylum policies, especially in light of increased asylum applications from those fleeing noncommunist states. Western governments have become tied up in legal and even constitutional knots over asylum claims, while providing costly relief assistance to refugees in distant countries. The international agencies mandated to cope with refugees and migrants are finding themselves stretched because of funding shortfalls, strained relief systems, and national challenges to their mandates. Meanwhile, much larger numbers of migrants are likely to move because of grim destitution.

A pressing issue for the next millennium will be balancing sovereignty concerns of states with the UN's role in protecting the basic human rights of refugees and economic migrants. Foreign policies, domestic politics, economic demands, and immigration laws have long been directly and indirectly linked and will continue to be so for the foreseeable future. Although immigration has become a major topic of domestic political debate in almost every Western country, it remains a subset of a much larger pattern of international migration that has only recently been addressed through multilateral diplomacy. Similarly, foreign assistance to refugees, while still provided through a combination of strategic as well as altruistic motives, is increasingly dependent on multilateral relief delivery efforts.

Global and regional accords dealing with refugees and economic migrants are helping to establish improved guidelines and procedures for the international community to deal with increasing numbers of people on the move. Even with such accords, though, governments will have to develop more sophisticated ways to manage migration flows within and across their borders. This might begin with the recognition that all migrations are a bundle of issues involving economic opportunities, political jurisdictions, ethnic tensions, human rights, basic needs, and environmental scarcity.

A political geographic perspective can demonstrate how these migration-influencing factors interrelate and thus can help policymakers better read the signposts that point to future international migration streams. Those governments able to meet this challenge will be the ones that address head-on the compounded problems of desperate migrants and unstable socioeconomic and geopolitical conditions. If they are to cope with projected international migration flows in a humane way, governments will need to act multilaterally and rely more on the expertise and operational capabilities of UN agencies and NGOs who are committed to working with migrants instead of against them.

Notes

1. See Everett Lee's classic essay "A Theory of Migration," *Demography* 3 (1966): 47–57.

2. The most cited geographic study of changing mobility patterns is probably Wilbur Zelinsky's "The Hypothesis of the Mobility Transition," *Geographical Review* 61 (1971): 219–249.

3. Perhaps the most influential researcher on economic aspects of rural-urban migration is Michael Todaro; a good summary of his early work on this is *Internal Migration in Developing Countries: A Review of Theory, Evidence, Methodology, and Research Priorities* (Geneva: ILO, 1976).

4. See numerous articles in the journal *International Migration Review.*

5. Stephen Castles and Mark J. Miller, *The Age of Migration: International Population Movements in the Modern World* (New York: Guilford Press, 1993). Also, Mary Kritz, *U.S. Immigration and Refugee Policy: Global and Domestic Issues* (Lexington, Mass.: Lexington Books, 1983).

6. See Ronald Skeldon, *Migration and Development: A Global Perspective* (New York: Longman, 1997). Lawrence Brown, *Place, Migration, and Development in the Third World* (London and New York: Routledge, 1991). Also, Philip Ogden's *Migration and Geographical Change* (Cambridge: Cambridge University Press, 1984).

7. See the landmark study by George Demko and Roland Fuchs, eds., *Population Distribution Policies in Developing Planning* (New York: UN Population Division, 1981).

8. See the essays in Richard Black and Vaughan Robinson, *Geography and Refugees: Patterns and Processes of Change* (London and New York: Belhaven Press, 1993). An earlier set of geographic perspectives on refugee problems is in John Rogge, ed., *Refugees: A Third World Dilemma* (Lanham, Md.: Rowman and Littlefield, 1987).

9. Michael Teitelbaum and Myron Weiner, eds., *Threatened Peoples, Threatened Borders: World Migration and U.S. Policy* (New York and London: W. W. Norton, 1995).

10. Demetrios Papademetriou, *Coming Together or Pulling Apart? The European Union's Struggle with Immigration and Asylum* (Washington, D.C.: Carnegie Endowment for International Peace, 1996). Also, William Brubaker, ed., *Immigration and the Politics of Citizenship in Europe and North America* (Lanham, Md.: University Press of America, 1989).

11. See Frank Bean, Georges Vernez, and Charles B. Keely, *Opening and Closing the Doors: Evaluating Immigration Reform and Control* (Santa Monica, Calif. and Washington, D.C.: RAND Corp. and Urban Institute, 1989).

12. Roberto Suro, *Strangers Among US: How Latino Immigration Is Transforming America* (New York: Alfred A. Knopf, 1998).

13. William Wood, "Forced Migration: Local Conflicts and International Dilemmas," *Annals of the Association of American Geographers* 84, no. 4 (1994): 607–634.

14. Nigel Harris, *The New Untouchables: Immigration and the New World Worker* (London and New York: I. B. Tauris, 1995).

15. Leon Gordenker, *Refugees in International Politics* (New York: Columbia University Press, 1987).

16. See UNHCR, *The State of the World's Refugees: In Search of Solutions* (New York: Oxford University Press, 1995); and U.S. Committee for Refugees, *World Refugee Survey, 1997* (Washington, D.C.: Immigration and Refugee Services of America, 1997).

17. William Wood, "From Humanitarian Relief to Humanitarian Intervention: Victims, Interveners, and Pillars," *Political Geography* 15, no. 8 (1996): 671–695.

18. Wood, "Forced Migration," pp. 607–634.

Chapter Nine

Exploiting, Conserving, and Preserving Natural Resources

SUSAN L. CUTTER

By the end of nature I do not mean the end of the world. The rain will still fall and the sun shine, although differently than before. When I say "nature," I mean a certain set of human ideas about the world and our place in it.... More and more frequently ... our sense of nature as eternal and separate is washed away, and we will see all too clearly what we have done.
— **Bill McKibben,** *The End of Nature*

Environmental degradation becomes an international, political, and geographic issue when it undermines a state's resource base and compromises its national security. Local resource issues can rapidly assume global significance as societies grapple with how best to use and sustain their cache of relatively scarce natural resources in relation to other countries. Disputes over natural resources, for example, often mar diplomatic relations. In fact, wars over resource disputes may be the rule rather than the exception in years to come, as countries attempt to reduce internal and external threats to their environmental security.

This chapter examines the international system of natural resource production and use. Natural resources are not evenly distributed on, below, or above the earth's surface. Some places are rich in highly valued natural resources, and other places are seemingly poor, leading to uneven access to

TABLE 9.1   Arable Land

	Percent of World's Land	Percent of World's Cropland	Percent of Region's Arable Land	Percent of World's Population	Per Capita Cropland (hectare per person)
Africa	22.6	12.9	6.3	12.7	0.27
North and Central America	16.6	18.7	12.5	7.9	0.61
South America	13.4	7.2	6.0	5.6	0.33
Asia	23.6	32.4	15.2	60.5	0.14
Europe	17.3	9.4	6.0	12.7	0.20
Oceania	6.4	3.6	6.1	0.5	1.86

SOURCE: World Resources Institute, *World Resources 1996–1997* (New York: Oxford University Press, 1996).

these basic building blocks of economic systems. Some resources are owned by a sovereign state; others are transnational in their distribution and must be shared. Still others are global and part of the earth's "common heritage." The use and management of these various types of resources are functions of political ideology, technology, and market forces, which, in turn, influence the type and rate of resource consumption. Variations in political ideology and economic affluence among countries and differences in access to technology and markets often lead to confrontations over the management of natural resources. These clashes over resources can occur at the local or national level, but disputes over transnational resources and the global commons are becoming increasingly frequent.

Diversity in Nature

Even before any human intervention occurs, there are fundamental inequities in natural resource distribution. Land areas vary greatly from country to country, as does the amount of arable land capable of supporting cultivated agriculture. Only 11 percent of the world's land is currently in cropland production, and this, too, is unevenly distributed (see Table 9.1). The rest of the land is either too dry, too cold, too wet, or too mountainous, or it has poor soils or is already severely degraded. Only 6 percent of Africa, for example, is arable, around 0.3 hectares per person (less than one acre). The largest percent of arable land is in Asia, a large, heavily populated region with a long tradition of intensive agricultural production. Although additional land can become productive through inputs such as water (irrigation), fertilizers, pesticides, and more drought- and pest-resistant crop varieties, there is still a finite amount of arable land. Perhaps more to the point, the economic cost and potential environmental problems increase as farmers are forced to use more marginal land (i.e., lands poorly suited for growing crops). Human population pressures, overgrazing, and the unsustainability of modern agricultural practices, which rely heavily on intensive fertilizer and pesticide inputs, are placing enormous burdens on land resources, especially in Asia and Africa. Although total agricultural outputs have increased worldwide since the 1970s, regional production trends are quite different. In Asia, food production has increased, along with population. In Africa, however, per capita agricultural production was stagnant during the last two decades, while population increased by 2.8 percent per year. Despite the global increases in food production, nearly one billion people are undernourished and go hungry every day.[1]

Water resources are also highly variable—bountiful in some regions and during some seasons, scarce in others. Societies routinely manipulate the supply and quality of freshwater resources. Dams and diversions capture, store, and channel freshwater for a variety of local, regional, and national needs—irrigation, power generation, industry, and domestic consumption. Because rivers can flow and aquifers can permeate through several countries, the potential for international disputes over their use is increased, especially when the water resources become scarce through natural variability or short-term droughts or through industrial or agricultural contamination.[2] Even when water is plentiful in supply, industrial and agricultural pollution may reduce the availability of drinking water from both freshwater and groundwater sources. The lack of safe drinking water is especially acute in the overcrowded urban slums and squatter settlements in developing countries, where diseases such as cholera and dysentery sporadically reach epidemic proportions. The lack of safe drinking water also affects many rural areas of the developing world.

Mineral and energy resources vary because of continental quirks in geology. Australia, Canada, Chile, China, Cuba, Russia, South Africa, and the United States have the largest reserves of nonfuel mineral resources (Figure 9.1). South Africa alone has 13.5 percent of the world's fifteen most important mineral reserves and a virtual monopoly on chromium and manganese (Table 9.2). Fuel resources derived from minerals (coal, petroleum, natural gas, uranium) are also spatially concentrated in North America, China, Russia, and the Middle East (Figure 9.2).

Uneven Access: Territorial Ownership Versus the Global Commons

Natural resources exist independently from human activity and are variable regardless of whether or not we choose to use them. Only when society finds some utility in some of the world's "neutral stuff" is a value given to the resource. Individual states have internationally recognized "rights" to

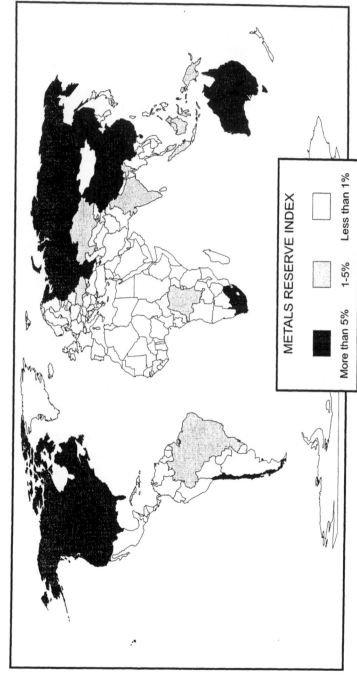

FIGURE 9.1 *Metals reserve index, 1996. The metals reserve index is the mean global share of fifteen minerals (copper, lead, tin, zinc, iron ore, manganese, nickel, chromium, cobalt, molybdenum, tungsten, vanadium, bauxite, titanium, and lithium) for each country. It provides a comparative measure of the location of the world's mineral powers. The percentages reflect that country's overall share of the world's known reserves.*

SOURCE: Data from the USGS Mineral Commodity Summaries.

TABLE 9.2   Major Mineral-Producing Countries

	Country	Percent of World Reserves	Percent of World Production, 1996
Bauxite	Australia[a]	28.2	39
	Guinea	21.1	13
	Jamaica	7.1	11
Chromium	South Africa[a]	73.3	42
	Kazakstan	4.3	20
	India	0.9	9
Cobalt	Zambia	6.0	25
	Canada	2.9	22
	Russia	2.6	15
	Zaire[a]	27.8	10
Copper	Chile[a]	26.7	28
	United States	14.8	18
	Canada	3.8	7
Iron ore	China	3.9	25
	Brazil	7.3	19
	Australia	13.9	14
	Russia[a]	18.1	8
Lead	Australia[a]	28.3	18
	China	9.2	16
	United States	16.6	15
Manganese	South Africa[a]	80.0	19
	Ukraine	10.4	15
	China	2.0	15
Molybdenum	United States[a]	45.0	46
	Chile	20.8	14
	China	8.3	14
Nickel	Russia	6.6	23
	Canada	12.7	18
	New Caledonia	13.6	11
	Cuba[a]	20.9	5
Tin	China	16.0	26
	Indonesia	8.2	20
	Peru	0.4	12
	Brazil[a]	25.0	9
Tungsten	China[a]	44.8	67
	Russia	11.9	18
	Canada	12.4	−
Zinc	Canada[a]	15.0	16
	China	3.6	14
	Australia	12.1	13
	United States	11.4	9

[a]Largest known reserves in the world.

SOURCE: World Resources Institute, *World Resources 1996–1997* (New York: Oxford University Press, 1996); USGS Mineral Commodity Summaries, 1997 (http://minerals.er.usgs.gov/minerals/pubs/stat).

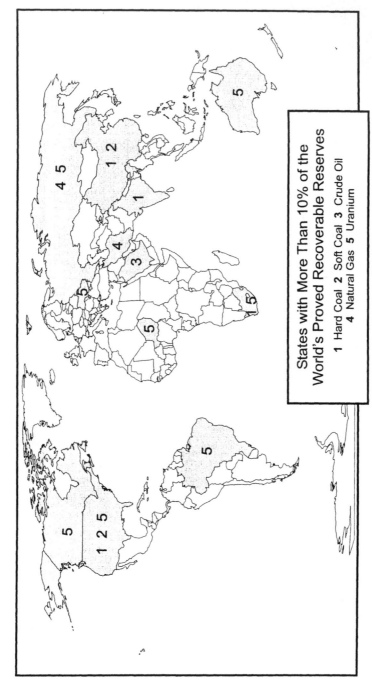

FIGURE 9.2 *Energy reserves. Countries with greater than 10 percent of the world's proved recoverable reserves for coal, oil, natural gas, and uranium are highlighted.*

SOURCE: Based on data from World Resources Institute, *World Resources 1996–1997* (New York: Oxford University Press, 1996).

natural resources that are part of their sovereign territory. This national control allows unlimited governmental access to these resources, which can then be sold to private corporations or developed by state-owned resource utilities such as PEMEX (the Mexican national oil company) or Hydro-Quebec (the provincial developer of hydroelectric resources in Canada).

Some resources, such as freshwater lakes and rivers, may transcend political or territorial boundaries, necessitating shared access. These transnational resources often require cooperative agreements between neighboring countries. Rivers, for example, may flow through many states and also provide territorial boundaries for some (such as the Rio Grande between the United States and Mexico). Diverting or reducing the flow of water upstream has considerable impact on downstream users. Dams such as Turkey's new Ataturk Dam near the headwaters of the Euphrates, which is to be used for hydroelectric power and irrigation schemes, may have considerable impact on the downstream users, Syria and Iraq, who fear a decline in water quality and quantity. Under a 1987 agreement, Turkey is to supply Syria and Iraq with a certain amount of flow, but there is no guarantee that Turkey will abide by this agreement should relationships between the countries sour as a result of Syrian support for Kurdish insurgents in eastern Turkey. Similar water-flow problems exist along the Colorado River and create a source of tension between the United States (the upstream user) and Mexico (the downstream user) and between California and Arizona.[3]

National boundaries are permeable to most environmental contaminants since pollution transported by air or water recognizes no nationality. The degradation of freshwater resources by pollution is a good example. Contamination of the Rhine from the 1986 Sandoz chemical spill in (upstream) Basel, Switzerland, affected not only Swiss residents but also residents in all those countries downstream (Germany, France, and the Netherlands). Transboundary industrial pollution has helped to strengthen the role of regional political alliances, such as the European Community, in regulating environmental quality. It has also fostered cooperation between environmental nongovernmental organizations (NGOs), who are increasingly adopting an international outlook.[4]

There are other resources, however, where ownership is less clear, such as the atmosphere, biotic resources in the deep oceans (such as whales or fish), or minerals in Antarctica. These common-property resources are extraterritorial and have been exploited by private individuals, corporations, or governments for their own profit because they have the technical and economic means to do so.[5] Since the exploiters are not held responsible for the resource, they have no vested interest in safeguarding it. This conflict between ownership, unequal access, and management responsibility is one of the underlying causes of the exploitation of the global commons. Common-property dilemmas are often solved by institutional arrangements, such as trade sanctions, multilateral treaties, or privatization. A variety of international treaties are now aimed at managing resources as part of the global commons (Table 9.3). Protection regimes such as the Conference on the International Trade in Endangered Species (CITES) and the International Whaling Commission have facilitated the recovery of certain commons. Yet some argue that these protection regimes can create an entirely new set of tragedies, especially for developing countries.[6]

The current debate surrounding greenhouse gas emissions is a good example of common-property dilemmas. The implementation of the Climate Change Treaty—including emissions limits for greenhouse gases and a carbon tax if voluntary reductions prove ineffective—represents an attempt at managing the most common property resource of all, the atmosphere. Because of the global impact of local and national resource-use policies, there is a growing recognition of the pressing need for international cooperation in protecting the planet. United Nations agencies as well as international nongovernmental organizations have played key roles in fostering multilateral treaties on resource use, illustrating the salience of global environmental politics.[7]

Controlling Interests

Technological innovations and economic motives provide access to resources, but ultimately it is political regimes that determine how resources are exploited, conserved, or preserved. *Exploitation* is the complete or maximum use of a resource for immediate societal gain or individual profit. *Conservation* is the attempt to balance current resource use with longer-term availability through efficient, prudent, and ecologically sound management. Finally, *preservation* is the nonuse of a resource—an insurance policy that guarantees full protection of the resource, leaving it unimpaired for future generations.[8] These strategies are employed by individuals, private corporations, and governments, and they vary according to changes in technological and economic conditions as well as political regimes.

Political Economy

Natural resources provide the raw materials for industrialization and economic development. The type and pace of development are determined by the political ideology of the ruling elites and a number of other factors. In capitalist-free market theory, resources are allocated in the marketplace where the producer and consumer (through the imperfectly understood principles of supply and demand) determine how much of a resource is

TABLE 9.3 Major Global Treaties on Natural Resources and Environmental Protection

	Date	Resource Protected	Number of Participating States[a]
Antarctic Treaty	1959/1980	environmental and marine resources	40
Nuclear Test Ban	1963	atmospheric protection from radioactive fallout	122
Ramsar	1971	wetlands, waterfowl habitat	61
World Heritage	1972	natural heritages	108
Ocean Dumping	1972	marine protection from dumping	72
CITES	1973	endangered species	109
MARPOL	1978	marine protection from oil spills/pollution	85
Long-Range Transboundary Air Pollution, Europe (LRTAP)	1979	regional atmospheric protection from transboundary industrial pollution	32
Migratory Species	1979	wild animals	46
Law of the Sea	1982	access to common-property marine resources	137
Ozone Layer	1985	ozone layer protection	68
Montreal Protocols	1987	atmospheric protection and CFC ban	69
Basel Convention	1989	protect land and water by restricting transboundary movement of hazardous waste	52
Bamako Convention, Africa	1991	transboundary hazardous waste movements	17
Climate Treaty	1992	reduction of atmospheric greenhouse emissions	161
Biodiversity Treaty	1992	protection of biotic resources, especially tropical rain forests	161

[a]Participating countries are either contractors or signatories to the agreements.

SOURCE: World Resources Institute, *World Resources 1992–1993* (New York: Oxford University Press, 1992); ENTRI data base from the Consortium for International Earth Science Information Network (CIESIN) (http://sedac.ciesin.org/entri/prod/charlotte).

used and at what price. The producer can be an individual entrepreneur, a private corporation, a multinational corporation (MNC), or a state.

In socialist systems, resources had no intrinsic value. Their value was derived from the labor that was used to produce them. The desirability of economic development and resource consumption reflected the goals and values of the people (embodied in the state apparatus), not those of special interests or elites. Instead of using the "free market" as the exchange between producers and consumers, there was a centralized economic management and distribution of resources by state-controlled producers.

Underlying both systems is a technocentric ideology—the mastery of and domination over nature.[9] Nature is viewed as a means to support increased material and economic growth. Domination and mastery over nature are achieved through science and technological innovations that not only improve our ability to exploit resources but also allow us to do so more quickly. Technocentrists feel that there are ample supplies of key resources and that scientific and technical advancements will alleviate potential resource problems, including dwindling supplies, without any decline in economic prosperity.

This worldview results in an exploitative approach toward resources, in which they have only one purpose—to serve humans.[10] Some argue that the ideology originates in Judeo-Christian theology; other scholars find more complex explanations for these exploitive attitudes.[11] They argue that "Western" societal processes—industrialization, urbanization, gender domination, capitalism, and scientism—provided cultural tools to dominate nature. Yet, capitalist societies are not the only ones to exploit and degrade resources. Examples of the socialist system's inability to handle the social costs of industrialization (rapid resource depletion and environmental degradation) can be seen throughout Eastern and Central Europe and in many of the republics in central Asia, regions long dominated by the former Soviet Union.[12] Regardless of how it is implemented (whether through capitalism, socialism, or some combination of both), technocentrism has become the dominant perspective on natural resource use since the industrial revolution.

Over the past thirty years, there has been a fundamental shift in attitudes about nature, moving toward a more ecocentric vision—living in harmony with nature. The accelerating pace of global environmental change and its impact on the human condition has led us to reconsider our basic relationship with nature and our role in transforming the earth.[13] Increasingly, the earth is viewed as a closed system with a finite supply of resources—the "spaceship earth" concept. Ecocentrism advocates preservation and strict resource conservation. Economic growth is tempered by resource stewardship to provide the basic needs of society but not the excesses of unbridled consumerism. Environmental protection and economic growth become in-

separable, with the latter tempered by environmental responsibility and social justice. Ecocentrism, the underlying principle of sustainable development and resource use, is not antiscience or antitechnology; rather, it emphasizes small-scale, appropriate technology to provide for basic needs.

Often these environmental ideologies are taken up by special-interest
groups who try to influence resource decisionmaking. Environmental activism or "green rage"—be it expressed through local grassroots movements (Green Belt, Chipko), environmental organizations (Greenpeace,
Earth First!), or formal political parties (Germany's *die Grunen*)—takes an
aggressive ecocentric stance in the protection of the environment. In contrast, multinational corporations and some governments often cling to
technocentric ideologies that form the basis for their management decisions. Governmental environmental agencies and more mainstream NGOs
stand somewhere in between, arguing for both stewardship of resources
and sustainable development of society—a pragmatic, conservationist
view.

Technological Change

Technology, the tool used to exploit resources, has, of course, changed dramatically during the past two thousand years. As technology has become
more complex and powerful, it has enabled modern societies to exploit resources more quickly and efficiently than ever before.[14] Technological innovations such as the steel plow and barbed wire helped transform the vast
prairies and grasslands of the U.S. Great Plains into one of the most productive agricultural regions in the world. Much earlier, the fifteenth- and
sixteen-century voyages of exploration and discovery—made possible by
technological breakthroughs in shipbuilding, mapping, and navigation—
profoundly altered the global patterns of food resources through "ecological imperialism" (the introduction of many species of plants and animals
from the New World to the Old World and vice versa).

But technological change and innovation have a price, often in the form
of unanticipated consequences on social, economic, and environmental systems.[15] The mass production of goods, for example, leads to mass inputs of
raw materials, which may require a global search for new raw materials
that results in their rapid exploitation. What begins as a solution to a particular scarcity issue at a specific time in a particular locale often leads to
greater resource destruction over larger areas and longer time periods.

Energy resources provide a good example. Currently, the world uses in
one year the equivalent amount of fossil fuel energy that it took nature one
million years to produce. In Europe after 1750, traditional fuel sources
(firewood, animal and plant waste, charcoal) were rapidly replaced by coal
as the industrial revolution progressed. In the 1880s, Welsh anthracite coal

was shipped worldwide on steamships until supplies dwindled. The need
for additional reserves of coal resulted in the transfer of mining technologies and operations to distant lands. Lower-grade coals were also mined at
greater depths, and surface deposits were strip-mined. The rapid exploitation of coal for fuel industrialization led to severe environmental degradation in many of the great coal-bearing and coal-burning regions in the
world: Russia, China, Australia, Western and Central Europe, and the eastern United States.

Although the first commercial oil well in the United States was drilled in
1859, oil did not become a competitor with coal as a primary fuel until the
internal combustion engine was invented. After that, however, oil use rose
exponentially, replacing coal as the primary fuel for heating and transportation by the middle of this century. Today, plastics, synthetic fibers,
pesticides and fertilizers, chemicals, and a host of other consumer products
are all manufactured from petroleum and its derivatives. Indeed, we are so
totally dependent on oil—from the clothes we wear to the food we eat—
that some observers refer to the latter half of this century as the era of the
"hydrocarbon society."[16] Many developments in international relations
over the past fifty years have been based on the worldwide demand for and
supply of oil.

Another technological innovation in energy production is the fission
process. First used to generate electricity in 1957, fission reactors made uranium a valued resource. Nuclear power currently accounts for 7 percent of
the world's total energy production and 70 percent of its electricity output.
But in several nations, nuclear power is the dominant electrical energy
source, including France (85 percent), Belgium (97 percent), Lithuania (97
percent), Japan (78 percent), and Korea (76 percent). Nuclear energy is a
relatively "clean" fuel; it does not contribute to acid rain nor does it increase the amount of greenhouse gases emitted into the atmosphere. However, a failure in this technology, as the Chernobyl meltdown aptly illustrated, can cause catastrophic environmental destruction at both the
regional and global level.

Market Imperfections

Even if countries have the desire and ability to exploit resources, they may
be unwilling to do so because it is unprofitable for them, especially within a
free-market economic system.[17] As the commodity becomes more (or less)
in demand, the price changes, which, in turn, affects future demand. A de-
posit of deep seabed manganese may be too expensive to exploit today, but
if technology allows cheaper exploitation, if new demands arise for its use,
or if prices rise substantially, it may become profitable to mine manganese
nodules in the future.

Scarcity affects supply, demand, and price relationships. If a natural resource becomes scarce while its demand remains constant, its value will increase. There are two types of scarcity—absolute and relative. Absolute scarcity occurs when supplies of a finite resource (minerals and fossil fuels) are insufficient to meet current as well as projected future demands. Relative scarcity occurs when there are short-term variations in supply as a result of natural hazards (frosts, floods, droughts) or the intentional manipulations of the market by resource producers.

Unfair competition exists in global markets based on the preferential trading policies of states or resource monopolies (cartels) that band together to gain an economic advantage by reducing production to keep prices artificially high. For example, between 1975 and 1981, the price of a barrel of oil rose from $13 to $35, largely because of the influence of the Organization of Petroleum Exporting Countries (OPEC). This thirteen-member cartel (consisting of Algeria, Ecuador, Gabon, Indonesia, Iran, Iraq, Kuwait, Libya, Nigeria, Qatar, Saudi Arabia, the United Arab Emirates, and Venezuela) controlled around one-third of the world's oil production during the 1980s. To reduce the influence of OPEC, many countries increased domestic production and reduced demand through greater efficiency, thereby improving their position on the supply-and-demand pendulum. The result is that oil prices now hover around $21 per barrel.

Oil dominates energy and economic markets, creating a "world order of oil." It strains the economies of both industrialized and developing countries who increasingly rely on imported sources and thus depend on the international marketplace to purchase their oil. Moreover, the fluctuating price of oil can cripple the national economies of both exporting and importing states. When oil prices are low, exporters (many of whom are single-commodity countries) suffer, and tensions between rich and poor producers over production quotas intensify (as they have within OPEC). When prices are high, importing countries suffer as their foreign debt increases and inflation accelerates at home. During the next two decades, the geopolitics of oil will still focus on the Middle East, a region with 75 percent of the world's proven reserves of oil.

Another factor in global markets is the role played by MNCs, private corporations that operate in several countries simultaneously. These corporations have the ability to shift production and marketing activities from one country to another, depending on where profits are greatest. Some MNCs (Shell, du Pont) are large enough to control and manipulate markets at the national level, if not globally. For example, more than half (56 percent) of the world's grain (between 1992 and 1994) was produced by only five countries—the United States (17 percent), France (3 percent), Russia (5 percent), China (21 percent), and India (11 percent), most of which was consumed locally.[18] A handful of private, family-dominated companies, transnational in their operations, dominate global trading in foodstuffs.[19] Collectively known as the grain merchants, they are primarily based in the United States and France and include the firms of Cargill, Continental, and Louis Dreyfus.

The economic diversification of MNCs and their ability to move commodities and money internationally limit the regulatory controls of individual governments. With a few exceptions, MNCs are the masters of the markets and the purveyors of technology to exploit natural resources and then deliver them to consumers. In seeking new resource reserves or to continue exploiting existing ones, MNCs must pay countries for access in the form of production rights or extraction fees. All too often, however, the prices are extremely low. The state consequently feels exploited but often lacks the technology to extract the resource itself. Internal political struggles and nationalistic fervor in many developing countries have fostered anti-MNC sentiment. The MNCs, viewed as a threat to national security, may then be seized by the host government. Iran's nationalization of Anglo-Iranian Oil (the forerunner to British Petroleum) in the 1950s and Chile's seizure of American copper companies during Salvador Allende's regime in the early 1970s are good examples. The internal political struggles of these states affect global markets, where matters of national security often collide with free-market competition and trade.

Affluence

Affluence is perhaps the greatest controlling factor in resource consumption as well as a harbinger for political confrontation over its use. The imbalance between the largest consuming nations and population size is a measure of this affluence. The United States, with slightly less than 5 percent of the world's population, consumes a disproportionate share of the world's resources. For many resources (arable land, timber, water) the United States is self-sufficient; it produces what it consumes. Japan, on the other hand, is relatively resource poor and must import most of the natural resources that fuel its affluence. For example, both the United States and Japan are the top consumers of many of the world's major minerals (Table 9.4); these resources are the essential building blocks of industrialized economies. The United States has considerable mineral reserves (copper, lead, tin, zinc, iron ore), but Japan's are extremely limited. Both countries rely (to differing degrees) on mineral imports to drive their industrial economies. In addition, the United States consumes 25 percent of the world's energy, 21 percent of it imported. Japan, which imports 87 percent of its energy supply, is more efficient in energy use, consuming only 5 percent of the world's supply with about 2 percent of the world's population.

TABLE 9.4   U.S. and Japanese Consumption of Mineral Resources in 1994 (percent of world's total)

	United States	*Japan*
Aluminum/bauxite	26.8	10.8
Cadmium	12.0	36.1
Copper	24.1	12.4
Iron ore	6.5	11.7
Lead	25.7	6.5
Nickel	15.6	20.1
Tin	15.4	13.6
Zinc	16.1	10.4

SOURCE: World Resources Institute, *World Resources 1996–1997* (New York: Oxford University Press, 1996).

Confronting Inequities: Armed Conflicts and Peaceful Compromises

Variations in ideology, technology, and economic capability, coupled with the unequal distribution of resources, set the stage for political confrontations over resources at the local, regional, and global scale. Regional inequities in wealth and resources within nations can precipitate civil unrest and fuel separatist feelings. The Nigerian Civil War (1967–1970) is a good example. The eastern region of Nigeria, rich in mineral resources (especially oil), was also the most developed and heavily populated by Ibos. When it declared its independence (renaming itself Biafra), a civil war escalated into an international issue as the central Nigerian government implemented a naval embargo to prevent oil exports. Moreover, since Shell and British Petroleum produced 85 percent of Nigeria's oil, the civil strife became a national security concern for the United Kingdom. France and the United States also had strategic interests in the region as well.

As societies have evolved, many have exhausted their own territorial *(in situ)* natural resources. Either at the state or village levels, perceived carrying capacities have been exceeded either by swift increase in population or by the rapid exploitation of resources and the subsequent degradation of the environment. The quest for land and the resources it contains has always been a motivating factor in national territorial expansions, as Ratzel suggested (see Chapter 2). To achieve economic and political dominance and internal environmental security, states have tried to expand their territory or their zones of influence. Some have resorted to military threats and actions to ensure their access to an adequate supply of natural resources.

TABLE 9.5   Selected Twentieth-Century Transnational Armed Conflicts over Natural Resources

	Countries/Conflict	Resource in Dispute
1932–1935	Paraguay-Bolivia (Chaco War)	oil
1967	Arab States–Israel (Six-Day War)	water
1969	El Salvador–Honduras (Soccer War)	arable land
1972–1973	Iceland–United Kingdom (Cod War)	fish
1974	China-Vietnam (Spratly Islands Dispute)	oil
1982	United Kingdom–Argentina (Falkland-Malvinas War)	fish, oil
1991	Iraq–United Nations Coalition (Persian Gulf War)	oil

SOURCE: A. H. Westing, ed., *Global Resources and International Conflict: Environmental Factors in Strategic Policy and Action* (New York: Oxford University Press, 1986).

A number of examples illustrate the transnational politics of resource control. The Iberian conquest of the Americas, for instance, was prompted by a need to find more resources, especially gold to replenish the Spanish treasury after its war with the Moors; the Spanish were much less excited about the scientific joy in discovering the "New World." World Wars I and II were partially caused by population pressures in Central Europe, providing a justification for German territorial expansion in the quest for more living space. Japan's expansionist tendencies in East and Southeast Asia were partially motivated by its own lack of indigenous natural resources.[20] Similarly, the Falklands-Malvinas War between Argentina and Great Britain was nominally fought over access to marine resources within the 200-mile exclusive economic zone (EEZ) that surrounds the islands. Domestic Argentine politics and national pride also played a major role in the armed conflict. Finally, the 1991 Persian Gulf War was partially fought over access to oil, one of the most important natural resources now influencing international relations. There are many more examples of armed conflicts between states over natural resources (Table 9.5).

Unrestricted access to natural resources continues to be an important element in foreign policy. For example, international disputes over water resources can mar bilateral relations between countries, often requiring regional intervention in the dispute.[21] Hydropolitics and unresolved water disputes could prompt the next regional-scale armed conflicts in Africa (control of the Nile), South Asia (dry-season flow in the Ganges), or the Middle East (reduced water flow and salinization in the Euphrates, Tigris, and Jordan Rivers).[22] The latter case is perhaps the most worrisome since water, not oil, is the most precious resource in the region.

Peaceful methods for resolving transnational resource conflicts are being tried by more and more countries who see the need to develop multilateral

TABLE 9.6   International Habitat Protection Systems

	Biosphere Reserves			*Wetlands*		
	Number	*Hectares (x 1000)*	*Percent of World Total*	*Number*	*Hectares (x 1000)*	*Percent of World Total*
Africa	44	23,198	10.7	54	4,400	10.2
Europe	127	88,767	40.8	411	6,914	16.0
North and Central America	74	36,928	17.0	61	14,467	33.4
South America	27	50,559	23.2	20	7,788	18.0
Asia	42	13,513	6.2	61	4,559	10.5
Oceania	13	4,745	2.2	45	5,139	11.9
World	327	217,710	100.0	652	43,267	100.0

SOURCE: World Resources Institute, *World Resources 1996–1997* (New York: Oxford University Press, 1996).

environmental alliances to ensure resource availability. One method to protect ecosystems is to create national parks around them. Globally, there are more than 9,500 areas covering 970 million hectares that are protected as national parks.[23] Ecuador is the international leader, with 15 protected areas covering 39 percent of its national territory. However, many fragile ecosystems straddle political boundaries, and bilateral and multilateral cooperation is needed to ensure the protection of these cross-border ecosystems. In 1988, six Central American countries banded together to establish a series of "peace parks." These parks are designed to preserve the region's disappearing rain forests—belatedly recognized as a precious global resources—and to promote sustainable development. In 1989, there were 68 border parks involving 66 countries.[24] Currently, international protection systems include Biosphere reserves (217 million hectares protected in 327 units), wetlands of international importance (43 million hectares in 652 units), and marine protected areas (covering more than 40 million hectares in 1,306 units). The latter are mostly located in Asia and Oceania. There is considerable geographic variation in international biodiversity protection (Table 9.6).

The End of Nature or a New Beginning?

The politics of natural resources are undergoing unprecedented changes as we near the end of the twentieth century. Bill McKibben's apocalyptic vision of the "end of nature" reflects the political realities of the 1980s—economic growth regardless of environmental cost.[25] Yet the political winds shifted during the late 1980s and early 1990s, for several reasons.

First, a sharp degradation of critical world resources was observed and measured, such as the expanding hole in the ozone layer. Second, the relative overconsumption and affluence of the few wealthy nations was widely publicized and criticized by developing countries for exacerbating the twin worldwide problems of resource abuse and environmental degradation. Continual degradation of the soil, water, and marine resources promises, ultimately, to reduce food production. Pollution in some regions has become so severe as to produce measurable public health impacts, even weakening the ability to bear healthy children. Ozone depletion, global climate change, and reductions in biodiversity increasingly threaten the basic life-support system of our species, creating new political realities in many parts of the world.

The rate of resource use, the scale of environmental degradation, and the looming exhaustibility of many of our most important resources are forcing governments to reevaluate their basic assumptions about economic growth and planetary stability. Many countries have realized that economic prosperity is inextricably linked to environmental health. Even the industrialized Group of Seven (G7—the United States, Canada, the United Kingdom, Japan, Germany, France, and Italy) have publicly acknowledged the importance of environmental issues in economic growth and global security. And although international security and global economic problems continue to dominate foreign relations, the international system of resource production and distribution is becoming a third arena of intense diplomacy.[26]

Conclusion

Environmental issues are at the center of many of the world's most pressing problems, but environmental protection strategies alone will not solve them. This became quite clear during the 1992 Earth Summit meetings in Rio de Janeiro, attended by 170 national delegations. "Sustainable development" became the rallying cry for many participants who viewed that as the only viable path states could take to ensure their own survival and environmental security. With sustainable development, resources are managed to provide for the needs of the present generations without compromising the ability of future generations to meet their own needs.[27] The implementation of sustainable development ideals, however, remains largely untested and requires both governmental intervention and locally based initiatives.

Within the United States, the public has slowly recognized the links between economic and environmental survival. Ideas about resource conservation that were viewed as radical a few years earlier (such as higher taxes for gasoline or consumption taxes on material goods) are no longer dismissed, mostly because they make economic sense. Implementation of these will require sacrifices, many of which government leaders are unable or unwilling to pursue. On the global stage, awareness is increasing as well, and the pace of conservation implementation seems a bit faster in many regions.

Future economic development will be tempered by environmental responsibility. Issues of poverty, land tenure, population, health, and women's status will be increasingly linked under the banner of environmental security. East-West tensions already have been replaced by strains in North-South relations, and they will intensify. The implementation of sustainable practices will mean dramatic changes in the quality of life for millions of people, both rich and poor. But the changes will not come easily, and they may result in altered political regimes, civil unrest, and realignments in foreign relations, all of which will further affect exploitation, conservation, and preservation of natural resources in the decades ahead.

Notes

1. World Resources Institute, *World Resources 1992–1993* (New York: Oxford University Press, 1992); L. R. Brown, "Facing the Prospect of Food Scarcity," in *State of the World 1997*, ed. L. R. Brown et al. (New York: Norton, 1997), pp. 23–41; G. Gardner, "Preserving Global Cropland," in *State of the World 1997*, ed. L. R. Brown et al. (New York: Norton, 1997), pp. 42–59.

2. S. Postel, *Dividing the Waters: Food Security, Ecosystem Health, and the New Politics of Scarcity*, Worldwatch Paper 132 (Washington D.C.: Worldwatch Institute, 1996).

3. M. Reisner, *Cadillac Desert: The American West and Its Disappearing Water* (New York: Viking/Penguin, 1986).

4. L. K. Caldwell, "Globalizing Environmentalism: Threshold of a New Phase in International Relations," in *American Environmentalism: The U.S. Environmental Movement, 1970–1990*, ed. R. Dunlap and A. G. Mertig (Philadelphia: Taylor and Francis, 1992).

5. G. Hardin, "The Tragedy of the Commons," *Science* 162 (1968): 1243–1248.

6. B. J. McCay and J. M. Acheson, eds., *The Question of the Commons* (Tucson: University of Arizona Press, 1987).

7. G. Porter and J. W. Brown, *Global Environmental Politics* (Boulder: Westview Press, 1996); S. Kamieniecki, ed., *Environmental Politics in the International Arena* (Albany: SUNY Press, 1993).

8. S. L. Cutter and W. H. Renwick, *Exploitation, Conservation, Preservation: A Geographic Perspective on Natural Resource Use*, 3rd. ed. (New York: John Wiley & Sons, 1998).

9. T. O'Riordan, *Environmentalism* (London: Pion, 1976); D. Pepper, *Modern Environmentalism* (London: Routledge, 1996).

10. L. White, Jr., "The Historical Roots of Our Environmental Crisis," *Science* 155 (1967): 1203–1207.

11. L. W. Moncrief, "The Cultural Basis of Our Environmental Crisis," *Science* 1709 (1970): 508–512; C. J. Glacken, *Traces on the Rhodian Shore: Nature and Culture in Western Thought from Ancient Times to the End of the Eighteenth Century* (Berkeley: University of California Press, 1967); C. Merchant, *The Death of*

Nature: Women, Ecology, and the Scientific Revolution (San Francisco: Harper and Row, 1980); C. Merchant, *Ecological Revolutions: Nature, Gender, and Science in New England* (Chapel Hill: University of North Carolina Press, 1989); J. Seager, *Earth Follies: Coming to Feminist Terms with the Global Environmental Crisis* (New York: Routledge, 1993).

12. M. Feshbach and A. Friendly, Jr., *Ecocide in the USSR: Health and Nature Under Siege* (New York: Basic Books, 1992); F. W. Carter and D. Turnock, eds., *Environmental Problems in Eastern Europe* (London: Routledge, 1993); J. Kasperson, R. E. Kasperson, and B. L. Turner II, *Regions at Risk* (Tokyo: United Nations University, 1995); D. J. Peterson, *Troubled Lands* (Boulder: Westview Press, 1993).

13. R. Carson, *Silent Spring* (Boston: Houghton Mifflin, 1962); W. L. Thomas, *Man's Role in Changing the Face of the Earth* (Chicago: University of Chicago Press, 1956); B. L. Turner II., W. C. Clark, E. W. Kates, J. F. Richards, J. T. Mathews, and W. B. Meyer, eds., *The Earth as Transformed by Human Action* (Cambridge: Cambridge University Press, 1990).

14. D. R. Headrick, "Technological Change," in *The Earth as Transformed by Human Action,* ed. B. L. Turner II, W. C. Clark, R. W. Kates, J. F. Richards, J. T. Mathews, and W. B. Meyer (Cambridge: Cambridge University Press, 1990), pp. 55–67.

15. L. Winner, *The Whale and the Reactor: A Search for Limits in an Age of High Technology* (Chicago: University of Chicago Press, 1986); E. Tenner, *Why Things Bite Back: Technology and the Revenge of Unintended Consequences* (New York: Alfred Knopf, 1996).

16. D. Yergin, *The Prize: The Epic Quest for Oil, Money, and Power* (New York: Simon and Schuster, 1991).

17. J. Rees, *Natural Resources: Allocation, Economics, and Policy* (London: Methuen, 1990).

18. World Resources Institute, *World Resources 1996–1997* (New York: Oxford University Press, 1996).

19. D. Morgan, *Merchants of Grain* (New York: Penguin, 1979).

20. A. H. Westing, ed., *Global Resources and International Conflict: Environmental Factors in Strategic Policy and Action* (New York: Oxford University Press, 1986); J. Clay, "Resource Wars: Nation and State Conflicts of the Twentieth Century," in *Who Pays the Price? The Sociocultural Context of Environmental Crisis,* ed. B. R. Johnston (Washington D.C.: Island Press, 1994).

21. M. Renner, *National Security: The Economic and Environmental Dimensions,* Worldwatch Paper 89 (Washington, D.C.: Worldwatch Institute, 1989); M. Renner, "Transforming Security," in *State of the World 1997,* ed. L. R. Brown et al. (New York: Norton, 1997), pp. 115–131; M. Renner, *Fighting for Survival: Environmental Decline, Social Conflict, and the New Age of Insecurity* (New York: Norton, 1996).

22. S. Postel, *Last Oasis: Facing Water Scarcity* (New York: Norton, 1992); P. H. Gleick, ed., *Water in Crisis: A Guide to the World's Fresh Water Resources* (New York: Oxford University Press, 1993).

23. World Resources Institute, *World Resources 1996–1997.*

24. Renner, "National Security."

25. Bill McKibben, *The End of Nature* (New York: Random House, 1989).

26. Porter and Brown, *Global Environmental Politics;* D. Deudney, "The Limits of Environmental Security," in *Flashpoints in Environmental Policymaking: Controversies in Achieving Sustainability,* ed. S. Kamieniecki, G. A. Gonzalez, and R. O. Vos (Albany: SUNY Press, 1997), pp. 281–310; S. Dalby, "Ecopolitical Discourse: 'Environmental Security' and Political Geography," *Progress in Human Geography* 16 (1997): 503–522.

27. World Commission on Environment and Development, *Our Common Future* (New York: Oxford University Press, 1987).

Chapter Ten

Geo-Analysis for the Next Century: New Data and Tools for Sustainable Development

WILLIAM B. WOOD

Earth Observation Revolution

For almost two thousand years—since Ptolemy mapped in his *Geography* a set of stimulating if inaccurate earth observations—geographers have been on a never-ending quest to better describe the known world. The desire to accurately depict the earth has been a consistent calling. This task has been difficult and exciting, marked by European expeditions to collect samples of new species, map unknown coastlines, and describe strange cultures in far-off lands. This first two-millennia era was characterized by painstaking, costly earth surface explorations. Although often heroic, these explorers were far from objective; their expeditions usually led to military conquest, socioeconomic domination, and radical transformation of the cultures and

---

The views in this article are those of the author and not necessarily those of the U.S. Government.

landscapes that had just been "discovered." Historically, earth observation and cartography were not exclusively scientific in their purpose, and they will not be so in the future (see Chapter 5).[1]

Exploration did contribute to an ever-improving geographic knowledge base that led to more accurate and scientific data collection. Museums are filled with collected samples from countless expeditions, and atlases are filled with the accumulated locations of places and features. And yet, as in other sciences, what is striking is not how much we know but how little. For example, despite the reams of maps drawn by generations of cartographers, most of the world remains unmapped at a scale, level of accuracy, and currency that would make our earth observations relevant for most potential users—which should be just about everyone. This ignorance of the earth surface is not just a problem of lacking good local data. At the global level we still do not know with any confidence the extent and condition of the world's threatened ecosystems, and we still do not know how to use existent data in a comprehensive and consistent way for managing sustained economic development.[2]

The international community is now poised for a crash course on earth observation that should have profound repercussions for environmental protection, natural resource management, urban planning, infrastructure investment, and other scientific activities in the next millennium.[3] We have already entered the next era of computer-aided, satellite-based earth observation. It is based largely on increasing numbers of small satellites—as well as continued reliance on airplanes—which serve as mobile platforms for collecting a wide variety of remote sensing data. These earth images form a "geospatial foundation" upon which field data can be georeferenced through the use of new computer-based geospatial tools, such as handheld global positioning system (GPS) receivers and PC-based geographic information system (GIS) software. These geospatial tools and data could help to fill in the wide gaps in our knowledge of the earth, just as ships, sextants, and chronometers served earlier generations of explorers. Although this earth observation revolution is transforming the way we collect geographic data from our "digital earth," the more difficult challenge may not be in the technological realm but rather in the institutional one. As during previous eras of exploration, earth science data collection can serve multiple purposes, noble and not so noble. The real challenge lies not with just taking more, sharper, and real-time digital images of our planet but with how we turn this huge stream of data into useful, accessible information that serves the elusive goal of sustainable development. This chapter reviews the political geography of applying geospatial tools and data to the most fundamental of all global issues: sound management of earth resources.

New Tools, Old Questions

A historical case study can shed light on the enormous impact of a single technological breakthrough in earth science data collection. In 1714 the British government passed the Longitude Act, which offered a generous award to anyone who could solve a problem of the highest national security importance: location.[4] The government offered this lucrative inducement because British seamen, who were dominating the world's seas, could already determine their latitude north or south of the equator by their knowledge of the stars, but their inability to figure out their precise east-west geolocation on a spinning globe was resulting in disastrous shipwrecks. What might seem an easy problem today was extremely difficult; many of the best scientific minds of the time offered solutions, usually based on extremely complicated star charts. A reclusive clock maker named John Harrison took up the challenge by offering a different approach to the problem, suggesting that what was needed was simply a clock accurate enough to keep perfect time for months and durable enough to withstand violent seas. For more than four decades Harrison designed and laboriously built several models of an intricate chronometer. When, after his death, the instrument was mass produced, it finally allowed mariners to accurately determine their longitude in reference to Greenwich mean time and enabled cartographers for the first time to accurately map distant coastlines. As Dava Sobel's chronicle of Harrison concludes, "He succeeded, against all odds, in using the fourth—temporal—dimension to link points on the three-dimensional globe."[5]

There are several present-day equivalents to Harrison's remarkable chronometer that are now creating bold possibilities for the emerging second era of remote sensing–based earth observations. Three geospatial tools (which enable referencing and organizing of information to specific locations and areas on the earth's surface) are described briefly here: satellite-based remote sensing platforms (RS), flexible and affordable GIS software, and increasingly portable and accurate GPS receivers. There are numerous technical studies of these three linked earth observation data collection, organization, analysis, and display systems. The focus in this chapter is not on how they work but on potential sustainable development decisionmaking applications (see Figure 10.1).

Earth Imaging

RS platforms vary greatly in what they do, but they all share one common function: They allow images of the earth surface to be taken from many hundreds of miles away in space. Most of us have seen the wondrous Landsat images, which since the early 1970s have produced multispectral im-

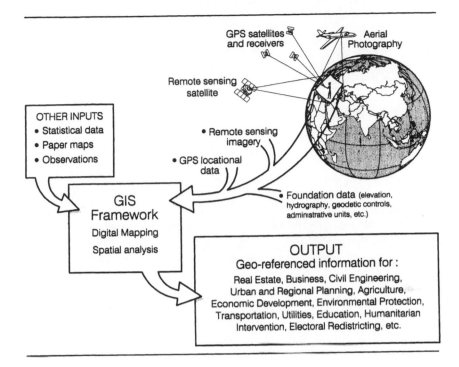

FIGURE 10.1   Geo-analytical tools, data, and output

ages—those that extend beyond the visible bands of electromagnetic energy. Landsat satellites, launched by National Aeronautic and Space Agency (NASA), generate images of about 30 meter resolution, which means that each pixel (the smallest unit of a digital image) covers about a football field on the ground. This resolution is fine for research on large areas, such as looking at seasonal vegetation changes for a whole country or extensive land use changes like rapid deforestation. Landsat is primarily a scientific tool, finely tuned for measuring major changes on the earth's surface, but its cost and scale have limited its broader application.

Since Landsat, other satellites have gone up with different "imaging packages." Some are part of weather and climate monitoring systems, the benefits of which are shown on the evening news weather forecast and in global climate change models. Several governments have invested heavily in sponsoring satellite imaging systems for both national security reconnaissance ("spy satellites") and potential commercial applications. European SPOT, for example, sells 10 meter panchromatic coverage (black and white images)—allowing for sharper pictures but less sensitive scientific measurement than Landsat. Russia, with its post–Cold War swords-into-plow-shares ethic, is now marketing high-resolution SPIN–2 film-based (versus digital) images that allow for very detailed coverage, but its package lacks real-time capability. The Canadian-based RADARSAT, launched in 1996, provides the opposite type of package—lower-resolution, but real-time radar-based images that can see through clouds night and day, a very useful tool if trying to quickly look at the extent of a natural disaster, such as an oil spill. India has also launched its IRS satellites, showing that such advanced technology undertakings are not just the domain of more developed countries.

That RS line-up is only the beginning. National and private satellite and other systems proposed for the next few years will generate a huge volume of earth observation data of unprecedented quality and, if the principles of supply and demand work as they should, affordability. Over the next few years, more than thirty land-observation satellites will cover swaths of the earth with imaging packages of varied frequency, resolution, and spectral characteristics. Major U.S. corporations committed to launching very-high-resolution (1+ meter) real-time systems aim to reach out to a potentially vast commercial marketplace where customers can buy and download customized imagery products over the Internet. As these satellite ventures compete for market share, they will be making new business alliances to make their earth observation data more user-friendly. Already, the groundwork for this has been laid with Internet-based data sharing and growing commercial applications of "digital earth" datasets.[6]

GIS Rules

A key factor in the marketing of these new remote sensing images will be how they are incorporated with other types of data to help customers solve a wide range of mundane problems—from routing delivery trucks to measuring soil erosion. GIS is a type of software that offers users a way of relating RS images with many other types of data collected within a common analytical framework. GIS is a key part of a geospatial methodology that organizes earth surface–relevant data—including locational data, areal patterns, and observed environmental changes—and allows for efficient analysis and visually compelling display of results. GIS software allows data "themes" (population, income, flora, etc.) to be viewed as "layers" that can be placed over each other in different combinations. Once available only on expensive workstations, GIS software now runs on laptop PCs and has become more "interoperable" with other types of software, thereby multiplying potential applications and users.[7]

Increasing flexibility, power, and affordability of GIS software has put it on the desktops of a wide variety of professionals: police detectives monitoring crime incidents; civil engineers working on sewage systems; tax officials reviewing property values; geologists trying to predict the intensity of the next earthquake; firefighters rushing to put out a spreading forest fire; environmental engineers measuring the spread of contaminants oozing from a hazardous waste site; and even politicians trying to gerrymander electoral districts. Federal, state, and local governments are using GIS tools to work more cooperatively on shared resource problems, such as in the Chesapeake Bay where a variety of pollutants coming from farms, poultry plants, and recreational use are affecting the viability of a vast ecosystem. Although almost all large city governments use some type of GIS package in at least one of their departments, a smaller group of farsighted city managers are using it as part of an "enterprise-wide GIS" that links all city departments together so that they can efficiently share common geospatial datasets.

At the international level, GIS is being used to look at environmental concerns that cross boundaries; the United States/Mexico Transboundary Mapping and GIS Initiative, for example, is undertaking a variety of imaging and mapping projects along the Rio Grande. The government of Qatar has developed the world's first national governmentwide GIS that ensures that all statistical data is georeferenced, spatially organized, and easily accessed by all official agencies. Intergovernmental agencies, both those working at the regional level as well as at the global level (such as the United Nations), are using GIS to look at datasets from many countries, such as national-level infant mortality rates, climate change observations, and worldwide deforestation.

Such varying types of data, though, need "ground rules" to be usefully applied within a GIS-based framework. Geospatial "foundation" datasets increasingly serve as the "base map" for a defined area, over which other, more transient types of data can be electronically laid down. Foundation data might include elevations, major natural features (rivers and coastlines), as well as some significant "cultural" features such as roads and international boundaries. As GIS becomes more widely used and as RS images proliferate, it will be increasingly critical that foundation data be accurately positioned on the earth surface and that RS images be aligned (or "rectified") to precise geodetic points. Such accurate georeferencing is an investment in a much larger geospatial infrastructure because, like the construction of a skyscraper, if the foundation is solid, many data stories can be added; if the foundation is flawed, the whole earth observation-based system will crack and become unreliable. If inaccuracy and confusion in the collection and use of geospatial data is to be minimized, agreed-upon standards and efficient means for sharing data need to be worked out among all potential data providers and users.[8]

Know Your Place

GPS is a Department of Defense–created and maintained navigational system that allows users to accurately locate themselves or other objects on the earth's surface. By triangulating signals from an array of twenty-four GPS satellites and through a variety of processing algorithms, users can "fix" their location with varying degrees of precision. The simplest, most widely available system allows users with a handheld GPS receiver to obtain locational accuracy within 100 meters, whereas the most restricted (by DOD), costly, and precise systems offer submeter accuracy. Like GIS, GPS has moved quickly from being available to a narrow group of technical users to being commonplace—they are now even being marketed to backpackers to track their treks.[9]

GPS, though satellite-based, is the direct complement to RS because it allows for efficient and accurate "ground truthing" of earth observation data. Field data collectors, ranging from utility workers to farmers, can combine their GPS-derived positions with GIS-based maps to create their own unique geospatial frameworks. They can then add quickly and easily new observation data tied to these accurately georeferenced points, creating a consistent basis for monitoring changes to the earth surface. GPS receivers are now available as part of digital cameras and survey equipment and are being "packaged" with GIS software to create a whole new seamless system for collecting geospatial data and tracking movement on the earth surface. Already car manufacturers have GPS as part of a built-in "trip mapping" application that allows drivers to plan their trips, to determine precisely where they are along their routes, and of course, to find how far they have to go.

Sustainable Development Challenge

The RS/GIS/GPS components of an evolving geospatial framework are synergistic in that they each contribute to an earth observation structure and process that is much greater than their individual parts (see Figure 10.1). What makes these sources and processes of geographic information of critical relevance now is that they offer a means for national governments and the international community to better respond to an enormous challenge: sustaining and even improving living conditions for the world's growing population while prudently managing the earth's depleted and degraded natural resources. At the very least, geographic information should help governments improve the way they work.[10]

Sustainable development has become an all-encompassing term for environmentally sound economic policies. Everyone lauds it as a goal, but achieving that idealized state is increasingly difficult. Governments face determined political pressure to encourage short-term economic returns over long-term sound resource management. Increasing numbers of young people in poor countries mean immediate demands for jobs, affordable housing, cheap food, public transport, urban infrastructure, and so on. In addition, decisions about use of national resource assets are often far from equitable; influential groups who control disproportionate amounts of a country's natural resource (private or public sector) have often been inclined to reap high profits now and worry about long-term environmental consequences later, much later. Finally, natural resources have traditionally been designated as either a "free" asset or seriously undervalued by economists and lending institutions, resulting in blatant resource waste, poor supervision of resource exports, and serious environmental contamination at extraction sites (see Chapter 9). Such conditions do not lend themselves to sustainable development implementation.

Environmentally Sound Development

Governments and international development organizations have recently begun to acknowledge the shortsightedness of policies that encourage uncontrolled resource exploitation. Apart from the international spotlight of the 1992 UN Conference on Environment and Development in Rio de Janeiro, there have been numerous global and regional fora focused on the dilemma of reconciling economic growth with environmental protection. The World Bank, bastion of prudent macroeconomics and past sponsor of mega-development projects, now argues that governments must rethink how they value their natural and human resources and, furthermore, orient their economic development policies around these reconfigured "national accounts." Likewise, the U.S. Agency for International Development's explicit mission is the promotion of sustainable development, as is that of the UN Development Programme.

Many of these international efforts are beginning to converge. A multi-agency conference in Bangkok, Thailand, in 1994 tackled the difficult issue of outlining "core data needs for environmental assessment and sustainable development strategies."[11] Participants disputed the "map myth" that suggests there is adequate and comprehensive cartographic coverage of the earth's surface—in reality, there are huge data gaps in what we know about the earth surface and a dearth of critical information required for sustainable development. To remedy this global ignorance, scientists in Bangkok recommended intensive collaborative work on ten high-priority core datasets that are deficient in most countries: land use/land cover, demographics, hydrology, infrastructure, climatology, topography, economy, soils, air quality, and water quality. The RS/GIS/GPS role will be central to this data collection effort. In addition, however, governments need to also work together to look at issues of geospatial data cost, security, and access, as well as promote capacity building in poor countries so that relevant sustainable development information—within a geographic framework—is put into the hands of those who need it most.

Collective Geographic Challenge—From Global to Local

International efforts to at least sustain viable living conditions for a growing population may be the ultimate challenge of the next millennium. Global climate change concerns have generated much academic and policy debate over rates of global warming, accuracy of data, and implications. Intergovernmental panels have looked at the available climatic evidence and issued cautionary warnings on likely results of continued increases in emissions of greenhouse gases. However, government capabilities to reduce greenhouse gas emissions vary, resulting in a complicated international implementation regime. Improved data on weather fluctuations, soil types, temperature trends, and hydrologic cycles can be used to determine likely climatic impacts on agricultural production. To be useful, though, such information needs to be spatially correlated, verifiable and reliable, and accessible to a broad range of sustainable development practitioners.

A giant step toward the goal of accurate worldwide environmental information is already being taken. The UN Food and Agricultural Organization, for example, continues to update its global soils database. The International Irrigation Management Institute and Utah State University announced in March 1997 a new World Water and Climate Atlas for Agriculture that pulls together in one database weather-related records from around the world from 1961 to 1990. The UN Environment Programme and the U.S. Geological Survey's EROS Data Center have developed Global Resource Information Database (GRID)–linked digital elevation data for the world and derived models showing drainage basins and watersheds. The EROS Data Center is also working on detailed vegetation maps (beginning with the United States) showing seasonal land covers, which combine AVHRR images with other environmental data.

Other efforts aim to improve access to such critical information. The Consortium for International Earth Science Information Network (CIESIN), for example, is using the Internet and data cataloging, search, and retrieval tools to link environmental and public health research agencies around the world. These decentralized electronic linkages foster information sharing and the creation of common metadata standards (information about the datasets—akin to a library catalog system) that could enhance international cooperation on sustainable development-related problems. CIESIN is one of forty-four World Data Centers established by the International Council of Scientific Unions and the only one to deal with the interface between environmental, demographic, and socioeconomic data.

Nongovernmental organizations (NGOs) have been increasingly active in using GIS-linked technologies to look at sustainable development-related problems and to protect threatened habitats (see Chapter 14). World Resources Institute (WRI), well-known for its worldwide datasets on natural resource inventories, is developing a "country data sampler" that compiles a digital atlas of population distribution at subnational levels. This work, already largely complete for Africa, directly complements other efforts that "map" ecosystems because sustainability can be assessed only by looking at the effects of population growth and land-use changes on a particular habitat. Other environmental NGOs, such as Conservation International, the Nature Conservancy, and the World Wildlife Fund, are using GIS to identify particularly fragile and/or biodiverse ecosystems, such as tropical forests, that are now being threatened. Once the true dimensions of these ecosystem threats are visualized with GIS tools and remote sensing data, these NGOs hope to use it to persuade governments to step up conservation measures.

The real benefit of GIS-based analysis of sustainable development, however, will be at subnational levels because that is where the most important decisions are made about land use. Anthropologists are using GIS to work with indigenous peoples to "map" their homelands so that they can then better argue their rights to these lands to government officials and help stave off land encroachment from outsiders claiming that no one has title to these lands and thus they are up for grabs (for farms, timber concessions, mines, etc.).[12] Land degradation poses another subnational and international dilemma. A recent study on Mexico used GIS analysis to show clear linkages between land degradation in semiarid and arid regions of Mexico, declining crop productivity and increasing rural poverty, and out-migration.[13] These land-use pressures will have profound economic implications for Mexico's 96 million people and even more for Mexico's projected population of 140 million in the year 2025. It will also greatly influence pressure on Mexicans to migrate to the United States over the next few decades. GIS-based analysis will not solve these very serious subnational and international problems, but it can shed light on the dimensions of resource degradation, help identify areas of greatest severity, and assist in planning land conservation and agrarian programs aimed at improving living conditions for rural Mexicans.

Spatial Data Infrastructures

The U.S. government is attempting to build a National Spatial Data Infrastructure (NSDI), which has the ambitious goal of ensuring that all geospatial data collected by government agencies are in a form that is compatible and thus shareable. The NSDI aims to establish a framework and set of standards that would allow all participating agencies to build up their geo-referenced datasets (which can range from county soil maps to census tracts) in a consistent and comprehensive manner—a clear cost benefit for all involved, yet difficult to systematically implement because of bureau-cratic inertia and project-driven budgets. The U.S. Federal Geographic Data Committee, chaired by the secretary of the interior, is tasked with try-ing to establish the rules and means for geographic information sharing among federal agencies and to support complementary efforts at state and local levels. The NSDI is thus the culmination of a political geographic process involving negotiations over the way geographic information is col-lected, organized, and provided to a wide variety of users.[14]

In the private sector, the Open GIS Consortium is working on common standards among GIS and remote sensing companies to ensure maximum "interoperability" of geospatial data, regardless of which software was used to produce it. Most recently, international discussions have begun for a Global Spatial Data Infrastructure (GSDI), which would help ensure that earth science information—wherever it might be collected and by whatever means—would be logically and systematically organized so that it could be readily understood by scientists, policymakers, or students around the world.[15] Achieving that global goal, however, will likely be even more diffi-cult than at the national level because of continued distrust among govern-ments over revealing possible "national security" secrets, releasing propri-etary spatial data, and conforming to a world standard. Like the biases built into cartographic products, GIS-derived products can be misused and misconstrued; they need to be constantly and carefully reviewed by geogra-phers of all countries to ensure that standards of accuracy and consistency are maintained.

An internationalized spatial data infrastructure is, of course, part of an abstract "global information infrastructure," which has profound implica-tions on international economic, political, and military relations. "Accu-rate, real-time, situational awareness is the key to reaching agreement within coalitions on what to do and is essential to the effective use of mili-tary forces, whatever their roles and missions."[16] Although "situational awareness" is rooted in geography, its application need not be only for "battlespace dominance"; such information has been also critical to coordi-nating international responses to humanitarian crises and environmental threats.

International efforts to improve the collection, analysis, and distribution of geographic information are fundamental to a larger, more difficult process of improving sustainable development decisionmaking. The process of including environmentally sound policy options, using appropriate, energy-saving technologies, and achieving concrete results at the community level has begun in the United States but will require a sustained political commitment—the challenge internationally is even more daunting and urgent. This larger effort is more problematic because it moves from the relatively objective world of collecting earth science and population data to the murkier realm of how such information is used, by whom, and for what purpose. This broader political geography of sustainable development has received much less interest and yet without an intense, pragmatic focus on improving natural resource management at local, national, and international levels, the rapidly growing volume of geospatial data will be largely wasted. The world cannot afford such disregard.

Conclusion

When Ptolemy developed cartography as a science to describe his world, it had, very roughly, about 100 million people; when John Harrison finished with his chronometer that allowed the world to be fully explored, it had about one billion people; by the year 2000 world population will exceed six billion and only twenty-five short years after that, it will exceed eight billion. More than 90 percent of that continued population growth will be in countries that already are poor and already have serious resource degradation and depletion concerns. The earth's surface as the resource base for rapid population growth, then, requires sound policies and management as never before. Improved resource management, in turn, must be based on more informed sustainable development decisionmaking. Although such information will by no means guarantee fair and wise decisions, it should make sustainable development programs more effective, provided that relevant information is appropriately shared. The RS/GIS/GPS triad of geospatial tools and the geographic knowledge derived from them has become a multibillion dollar industry. The volume and quality of geospatial data will continue to increase, but the hardest part may be to take all that data and use it to better inform sustainable development decisionmaking, whether at a town hall meeting or the UN General Assembly.

Data alone, even if in the form of spectacular looking images of the earth, do not solve any development problems. It is easy to oversell the power of geospatial data as a development panacea, when in fact it is just a means of capturing particular types of earth surface information and a way for that information to be effectively presented for decisionmakers. The political geographic reality in which GIS-derived information comes in to play must be recognized: Critical resource decisions that affect much of the world's population often are neither democratic nor farsighted.[17] With that caveat in mind, however, geospatial data has much to offer for those who have the wisdom and means to use it. Governments working with the private sector and at both subnational and international levels must make important decisions about the use of geospatial tools and data over the next few years of the new millennium: how much to invest in them and for what types of products; who will have access to them, in what time frame, and at what price; and, most importantly, how to use them appropriately to improve living conditions.[18] Fundamental political geography issues, then, will revolve around our collective use of these powerful new tools and the growing "virtual library" of geospatial data. Perhaps not unlike Ptolemy two millennia ago, we will still be wondering how to use our new observations to better understand our world.

Notes

1. See Norman Thrower, *Maps and Civilization* (Chicago: University of Chicago Press, 1996); and John Noble Wilford, *The Mapmakers: The Story of the Great Pioneers in Cartography from Antiquity to the Space Age* (New York: Vintage Books, 1982).

2. See John E. Estes and Wayne Mooneyhan, "Of Maps and Myths," *Photogrammetric Engineering and Remote Sensing* 60 (May 1994): 517–524.

3. See Stephen Hall, *Mapping the Next Millennium* (New York: Vintage Books, 1992).

4. See Dava Sobel, *Longitude: The True Story of a Lone Genius Who Solved the Greatest Scientific Problem of His Time* (New York: Walker, 1995).

5. Ibid., p. 175.

6. See Office of Technology Assessment, U.S. Congress, *Remotely Sensed Data: Technology, Management, and Markets* (Washington, D.C.: U.S. Government Printing Office, 1994). Also, William Stoney and John Hughes, "A New Space Race Is On!" *GIS World* 2 (March 1988): 44–46.

7. For overviews of GIS applications, see David Martin, *Geographic Information Systems: Socioeconomic Applications* (London and New York: Routledge, 1996); and J. L. Star, J. E. Estes, and K. C. McGwire, eds., *Integrating Geographic Information Systems and Remote Sensing* (New York: Cambridge University Press, 1997). Also see numerous articles of specific GIS-related projects and new technologies reported in *GIS World* and *Geo Info Systems* magazines.

8. For a thorough discussion of institutional constraints and opportunities for working with GIS-based data, see Harlan Onsrud and Gerard Rushton, eds., *Sharing Geographic Information* (New Brunswick, N. J.: Center for Urban Policy Research, 1995).

9. See discussion by the National Research Council, *The Global Positioning System: A Shared National Asset* (Washington, D.C.: National Academy Press, 1995). For discussions of various GPS and GPS/GIS applications, see numerous articles in *GPS World* magazine.

10. See discussion and recommendations made by the National Academy of Public Administration, *Geographic Information for the 21st Century: Building a Strategy for the Nation* (Washington, D.C.: National Academy of Public Administration, 1998).

11. See UN Development Programme and UN Environment Programme, *International Symposium on Core Data Needs for Environmental Assessment and Sustainable Development Strategies*, vols. 1 and 2, Bangkok, Thailand, November 15–18, 1994. For a good introduction to the use of geospatial tools for resource management, see V. Alaric Sample, ed., *Remote Sensing and GIS in Ecosystem Management* (Washington, D.C.: Island Press, 1994).

12. See articles in *Cultural Survival Quarterly,* Winter 1995.

13. See the report by the Natural Heritage Institute, *Environmental Degradation and Migration: The U.S./Mexico Case Study* (San Francisco: Natural Heritage Institute, 1997).

14. See the Federal Geographic Data Committee report, *Framework Introduction and Guide: National Spatial Data Infrastructure*, U.S. Geological Survey, 1977.

15. See William B. Wood, "A Jeffersonian Vision for Mapping the World," *Issues in Science and Technology* (Fall 1997): 81–86.

16. See Joseph Nye and William Owens, "America's Information Edge," *Foreign Affairs* 75 (March-April 1996): 20–36.

17. See John Pickles, ed., *The Social Implications of Geographic Information Systems* (New York: Guilford Press, 1995).

18. See the report by the Mapping Science Committee of the National Research Council, *The Future of Spatial Data and Society* (Washington, D.C.: National Academy Press, 1997).

Part Three

International Processes of Geopolitical Change

Chapter Eleven

People Together, Yet Apart: Rethinking Territory, Sovereignty, and Identities

DAVID B. KNIGHT

Human *communities* are not discernible by the human eye from an earth-orbiting spacecraft, but many features, some of human origin, can be seen. A person on board a spacecraft can identify major physical features (such as Madagascar, South Asia, the Iberian Peninsula, Australia, and perhaps New Zealand) and may reflect on the thought that millions of people live "down there." Whatever can be seen will not in itself reveal *who* lives below nor anything about the nature of their communities or socioeconomic and political organizations. Some straight lines on the land (due to sharply contrasting vegetation types on one side as compared to the other, or significant breaks in land-use patterns) may suggest the existence of political boundaries (as across parts of western Canada and the United States), but otherwise it is not possible to discern from space much of any partitioning on the earth's surface nor, more especially, by what means people form communities and relate to and structure political territory. The cliché "spaceship earth" suggests oneness. Once on the surface, however, the reality is otherwise in numerous respects.

Imagine the spacecraft descending: At what moment would one see discrete settlements, which might, just might, suggest more clearly how people and their territories are divided? In posing this question, there is an important assumption, that is, that divisions exist! The late Boyd C. Shafer concluded that humankind is more alike than not, yet he spent his life studying what keeps people together and yet apart, namely, nationalism.[1] People together, yet apart! This chapter considers this thought, from a politico-geographic perspective.

Basic Attributes of the State

The basis of international society today is the territorial partitioning of the earth's surface into States and the application, or denial, of sovereignty and some related attributes to the resulting territorial units, all of which are set within an international system of States.[2] There has never been a fixed number of States although there have been periods of reasonable stability mixed with other periods when the number of States has changed markedly. Four periods of change stand out this century: following World War I, with a significant impact on the map of Europe; during and following World War II; the 1960s decolonization drive in various parts of the Third World, most notably in Africa; and, in 1989 and thereafter, the disintegration of the former USSR and some of its satellite States in Eastern Europe (including Czechoslovakia), plus politico-territorial and other changes in Yugoslavia and in various parts of Africa and Asia, including the transfer of Hong Kong to China by the United Kingdom.

Some common criteria must be met before a State can be said to exist, although the particular reasons why States exist varies quite markedly. The criteria for Statehood include the following: a particular defined territory; a permanent resident population; a constituted effective government; formal and real independence; sovereignty; recognition by other States in the international system of States; the expectation of permanence; the capacity to enter into relations with other States; a State apparatus; a circulation system; an organized economy; and various "fictional parts" of States, such as official residences of foreign diplomatic envoys.[3] Of these several criteria, territory, population (especially when considered from the perspective of group identities), and sovereignty are fundamental, and so they form the focus of this chapter even as it is acknowledged that the other criteria are important in the overall scheme of things.

The concepts considered in this chapter are human constructs: They are not givens. They are linked, each one to the others, in myriad changing ways, hence it is somewhat artificial to separate them. Even so, it is useful, at the outset, to separately identify some largely legal dimensions before reconsidering them in a more conceptual and interrelated manner, in conjunction with some additional concepts.

Territory

From the perspective of international political geography—and international law—any State needs a particular defined territory that is not shared with other States (although other States, or minorities in other States, may covet part or all of the territory) and that has limits that are more or less clearly defined (even if some of the boundaries may be unexplained or in contention). Territorial exclusivity pertains to the particular land area, its offshore waters and portions of rivers and lakes along or through which international boundaries exist, airspace above the areal limits of the territory (although the height of control in such airspace is contingent upon means for controlling entry into that space), and, obviously, subterranean areas (including the resources therein) and rivers, lakes, and so on that are internal to the territory. The essential component is land, to which all other attributes are ancillary. Oppenheim pithily summed up the essential importance of territory: "A State without territory is not possible."[4]

When a territory is delimited and organized as a State, there is a juxtaposition between jurisdiction and territory; the territory becomes infused with a legal function, a function that coincides with the existence of the State within the international system of States. Further, the State's territory provides both the base and the frame for the exercise of power, at Statewide and lesser (sub-State) scales, and indeed, as Sack observed, "the State is reified by placing it in space."[5]

A State's territory, in terms of acquisition, size, and shape, may vary from place to place and, sometimes, through time.[6] Competing claims to territory are often the basis for conflict, for "others" may claim "mine," and vice versa. Too many people have died due to territorial conflicts. International society has in recent times reached new levels of understanding and cooperation, yet there is still no basis for believing that conflicts over territory have necessarily ceased. Indeed, boundary disputes—varyingly at the (greater or lesser) margins of territory—form the major reason for conflict,[7] with all sorts of reasons being cited for territorial claims being made.[8] Most conflict over territory is based on some (sometimes quite dubious) historical claim.

With respect to "historical" claims, a problem remains, namely, how far back in time can contemporary claims be based. Consider, for instance, Israel's biblically based claim to Palestine and Indonesia's "precolonial" claim to East Timor. These cases raise related issues that arise from the fact that almost all of the earth is now claimed by existing States. Any expansion of current States and the creation of new States will necessarily involve existing States' territories. This clearly means that the territorial integrity of some existing States will someday be threatened. Since territorial integrity of States is believed to be a fundamental quality of Statehood, violent reaction can be expected by threatened States, although recent experience has demonstrated that violence need not always occur.

Territory is *the* critical quality for Statehood. It is a legal concept, and yet it is also much more. Territory without people is meaningless, for, indeed, as noted above, without people there is no territory—it is a human construct.

Territory alone does not constitute Statehood; but Statehood, first and foremost, is tied to territory. Territory is at once the territorial frame for certain
types of human societies and the substance upon which they live, and also
the essential basis for international organization, as currently structured.

Population

For a State to exist there must be a permanent resident population. The
government speaks and acts (or claims to speak and act) on behalf of the
total population of the State. The term population is neutral; other more
emotive terms are often used to describe a State's population. Primacy is
given to those who automatically acquire nationality (through birth) and to
those to whom nationality is otherwise granted. Nationality is dependent
upon Statehood, for without Statehood nationality would not exist. It is
generally accepted that the population of a State will thus have a sense of
nationhood but also, as a nation, that it will use its nationalism in their
search for national stability and security.

Nationality with respect to individuals refers to the quality of being a
member or "subject" of a specific State. Each State is free to decide who are
nationals and how nationality may be granted or rescinded. Some States
automatically withdraw nationality if, for instance, a new nationality is acquired elsewhere by naturalization or marriage. If the withdrawal of someone's nationality is based on political grounds then it will likely be condemned internationally, for such an act is commonly regarded as being of
doubtful validity in international law. Of course, nationality can be "returned" by governmental or legal action—as with the case of some now famous exiles from the former USSR. In contrast, some States hold that nationality can never be "lost," hence the slogan "once British, always
British!" Some States now permit dual citizenship.[9]

"People" is another word often used with reference to a State's population. This word has international legal meaning since, for instance, the UN
Charter and numerous other legal instruments refer to "all people" having
the right to self-determination. However, when used in that sense there is a
debate: Does the use of the word people refer only to *all* of the people of a
State, regardless of whether or not they all relate to and identify with the
national identity and the national government? Some scholars hold that the
word people can also refer to *parts* of the national population in *parts* of
the State. The latter people, as a minority (e.g., some Quebeckers within
Canada) or even as a dominated majority (e.g., Black South Africans during the *apartheid* period), may have primary allegiance to a sub-State "national" identity (or identities), which thus separates them from the remainder of the population of the existing State. Despite the very gradual
acceptance by some States of the term "people" to refer to some sub-State

groups, primacy is still retained for the term being applied to the whole "people" of a State, not to minorities within it. Indeed, there is a paramountcy given under international law to the population of the *total* State over *parts* of either the population or territory within.

This is not to suggest that international law ignores other measures of humanness. For instance, there is international legal recognition of certain "minorities"—as individuals and as collectives—within States through such human rights instruments as the UN Covenant of Civil and Political Rights, the UN Covenant of Economic, Social and Cultural Rights, and the Helsinki Final Act.[10] But, and here we raise an important point about perception and self-definition, some such minorities—specifically, many indigenous peoples in various States—do not see themselves as mere "minorities," for they know they are "people" and thus are due international recognition!

It should be clear that mention of a State's permanent population, or people, raises questions pertaining to how group identities are defined and how they relate to the State. Group identities have religious, political, cultural, historical, and psychological bases. For our purposes it will be enough to refer to group territorial identities, a term generated to include terms such as tribe, ethnicity, ethnonationalism, nation, nationalism, mininationalism, sub-State nationalism, and so on, all of which are in the literature.[11] All of these group identifiers ultimately refer to a group's distinct character that sets it apart from others and which may have a reasonably observable territorial dimension, being dominantly or primarily located in a particular portion of a State (as with Scots in the northern part of the United Kingdom, Basques in Euskadi in northern Spain, and Sikhs in the Punjab in India).

Although the *ideal* of achieving a good areal fit between "nation" and "State" leads some "nations" to seek the creation of their particular "nation-States," the reality is that the concept of nation-State is questionable and so the term is of limited value.[12] Regrettably, the term "nation-State" is used by political geographers and others when what is meant is either, separately, "nation" or, more generally, "State." The word "nation" is too often carelessly and incorrectly used to refer to "States." Also, the term "nation" is all too often used when the more accurate term "national society" would be nearer to the truth.

Some States give shelter to refugees. The latter, by definition, are "homeless" and are said to be Stateless; they lack both national and international identity. This harsh reality, and the hesitation of States to fully support the right to a nationality, means that most refugees will continue to lack nationality due to their forced removal from their respective homelands, their States of birth. In contrast to refugees who have been forced to leave their homes due to war, famine, natural catastrophes, and so on, there are individuals who may have left their States under duress but who are not seen to have given up their nationality. Political exiles may manage to win recognition, loyalty, and support for their causes amongst people and even governments in other States.[13]

Sovereignty

Sovereignty has in the past one hundred years come to apply as a legal presumption only to territories formally constituted, accepted, and recognized as States by other States in the international system of States. Sovereignty, or the unqualified competence that States prima facie possess, implies competency to control the territory and its contents and also relationships with other States through the totality of powers that States, under international law, have and may use. Sovereignty once was equated with the monarch or with a group of people who exercised rulership, but it now pertains to an impersonal and legal prescription. Today, it is the State that wields sovereignty, and the exercise of a State's authority over its territory implies that sovereignty is complete and exclusive. A sovereign State is under no obligation to accept people, goods, and ideas from other States, and it may restrict or otherwise control all that is within the territory.

Sovereignty implies recognition by other States in the international system of States. Recognition, in turn, implies that each State respects the rights of other States. Recognition, as used here, refers to the recognition of the entity, the State, and not necessarily also the government. A government can be rejected by other States—perhaps with sanctions being imposed against the State because of that government's actions—or a government may fall even as the State continues to exist. Thus, for example, when Lebanon was in tatters in the 1980s and there was no effective government, the State was still said to be in existence. However, for a *new* State to be given recognition, a constituted and (reasonably) effective government has to exist (as in the cases of the new States that emerged from the recently failed USSR), and there has to be at least the promise of permanence.

Sovereignty becomes "legal fiction" if a State loses effective control of its territory (perhaps by the occupation of it by another State) and will be discarded by other States if it becomes clear that there is no prospect for that sovereignty to be regained. Finally, sovereignty cannot reside in a "government-in-exile," for sovereignty, under international law, implies actual "possession of, and control over, a territory."[14]

Elaborations and Complications

With the above thoughts in mind, it is instructive to reconsider the three basic terms from a more conceptual perspective and to discuss them with an awareness of how they interrelate among themselves and with some additional concepts.

Territory

Human societies create "territory" out of meaningless "space," for it is human societies that partition space and use it for their betterment. The delimited territory thus is special for the people concerned, especially when it is acknowledged as being sovereign territory within the confines of an independent State. A territory is by definition separate from others' territories, not only physically but generally also ideologically, for the people involved will have a territorial ideology that reflects their understanding of themselves in relation to their own territory and that helps guide, if not necessarily governs, the relations the State will have with other States.[15]

A delimited State-territory is, of course, but a particular type of region on the world's political map; any State covers only a particular portion of the surface of the earth. As with all regions that pertain to human social organization, a State's territory is a social construct.[16] The areal dimensions of territory also clearly are not givens, for it takes people to decide where the bounds of territory should be, and to change them or leave them alone once they are agreed upon. After all, "territory is not; it becomes, for territory itself is passive, and it is human beliefs and actions that give territory meaning."[17] The bounds of territory can simply be delimited (i.e., agreed to, perhaps by treaty, and thus, generally, are written in some fashion), but they may also be demarcated (i.e., be physically marked on the land).

Territory is real in the sense that it can be seen, felt, (sometimes) smelled, walked on, flown over, manipulated, and thus altered. It is, in this physical sense, quite concrete, for it has substance, can be measured, and can be changed (as with the building of new highways, the spread of towns, the changing of the land tenure system, etc.). The internal characteristics of the territory will thus change in the passage of time, as the society and its economy develop. The external dimensions of the territory may be altered too, but for this to happen there will, necessarily, be a change in the territory's relationship with another territory or other territories.

But territory is more than just a physical and measurable entity, for it is also something of the mind and people impute meaning to, and gain meaning from, territory. Indeed, many people fully believe in the landscape of "their" territory as a living entity that is filled with meaning. Such beliefs are psychologically and culturally based and therefore exist, at one key level, simply as parts of the "geographies of the mind." But since people's cultural ecology and spatial patterning (as in agricultural and settlement systems) can be powerfully influenced by the people's beliefs, it is, in turn, often possible to "read" them from the landscape. They can be inferred by the creation of landmarks (personal ones and, of significance here, major ones, such as monuments and shrines, which may have "national" significance);[18] from the naming of places;[19] from the "sacredness" invoked with respect to both specific parts or the generalized "national" whole of the territory;[20] and so on. The territory and the nation are celebrated as providers of security via the singing of a national anthem and the acknowledgment of a national flag. Above all, since the territory "is the very basis on which national existence rests, *true* citizens will be prepared to give their lives" in defense of "the 'sacred soil.'"[21]

Territory, at one and the same time, will serve the State as the locus of an inwardness, apart from international links, and yet also as the basis for linking with people and territories elsewhere. Political geographer Jean Gottmann has suggested that a State's territory—as "the model compartment of space resulting from partitioning, diversification, and organization"—offers a people dichotomous options: security (to look inward to preserve the integrity of the society against outside forces) and opportunity (from which to reach out to other societies). It is useful to think of these competing options as being at opposite ends of a continuum and to be aware that a State may tend more toward one end of the continuum than the other in its trade and foreign policies and actions. However, a change of stress from one to the other of these options may occur, as when a State, after a period of stressing opportunity by means of open international trade, takes a more isolationist stance and imposes tariff barriers to "protect" home producers and markets. Clearly, from this perspective, involving opportunity versus security, the concept of territory has implicit in it the possibility for conflict.

Territory may also form the basis for conflict in a different sense, as when competing claims are made with respect to a particular territory. Such claims need not involve two States, for competing claims can exist within a State's territory. Consider the following definition of territory: "space to which identity is attached by a distinctive group who hold or covet that territory and who desire to have full control over it for the group's benefit."[22] This is a provocative definition in that it links territory with people (who have distinct group identities) and the desire for control—that is, self-control by the people who inhabit the territory. Clearly, territory cannot be considered, from a political geographic perspective, without reference to the concepts of identity and control. Attention must therefore be placed on whether "people" with distinct identities have control of themselves within their own territories.

Although a government will normally represent the majority of the people within the State, there are many situations in which such does not pertain, as in Fiji (where a racial/"traditional" minority claimed to speak for the whole "people" of the State, when in fact it did not), South Africa and

Guatemala (where minorities ruled), China (where power and authority is maintained by a minority through the threat and use of force), and, of quite different types, Israel/Palestine (where the Government of Israel, despite notable attempts to effect peace and so permit change, continues to prevent the full expression of self-determination for the Arab majority in the Occupied Territories due to the continuing threat of violence and the concern for secure borders) and East Timor (where the people in an illegally invaded territory have suffered dreadful persecution and death even as the world turns a blind eye—this in contrast to the dramatic military response by some of the world's States to Iraq's invasion of Kuwait). A scale problem exists: For instance, the Chinese minority in Tibet dominate the majority, Tibetans, who are, in turn a minority in China.

How can minorities react? At a minimum, the minorities (or minority-dominated majorities) may seek appropriate and adequate recognition. But who is to decide what "appropriate and adequate recognition" means? Most such peoples want to gain, at minimum, the right to live their lives peacefully, without what they perceive to be the destructive power of the State taking and damaging their land and resources or otherwise influencing their lives. These thoughts turn us again to consider the degree to which a State's population can be regarded as having some linking "national" identity.

It may be that the total population of a State draws meaning and strength from a collective sense of self as a nation. Nationhood is not a necessary attribute of Statehood, however. Even so, the people of the State (especially those born therein), who may indeed see themselves as forming a nation, will draw strength from "their" territory. Of the latter, as Shafer noted, "in diverse ways [they will] love it and oppose any diminution of its size."[23] They may have a nationalism that has a real or created historical basis and is tied to the territory occupied by the State—and, perhaps, coveted territory in another State's control. A nationalist ideology will be formulated to interpret the occupation and control of the territory, both in the past and as a plan for the future. The ideology is based in part on a distinctively created "iconography" that reflects the set of symbols in which the people of the State believe and with which they identify.

The "Janus-like quality, [of] looking both ways to the past and the future, is equally relevant for States which seek to create and reinforce a sense of nationhood, and for nationalisms that oppose existing States in their attempt to carve out autonomous identities."[24] This last observation refers to the many instances where some inhabitants of certain States have a *sub*-State "national" identity, that is, an identity that pertains to only part of the total population of the State within part of the total territory, who may seek to have that identity more fully or totally recognized, within the State and perhaps even internationally.[25] If the latter is desired they may have to resort to secession, as part of the population in part of the State break-off to become a "nation" in its own new State, with a new nationality thus being created. But secession presents a threat to the international system inasmuch as it represents a challenge and threat to an international order that is based on the concept of the sovereign State.

As noted, the territorial integrity of States is commonly held to be more important than any minority claims for self-determination from within a territory. Thus, for instance, the Aland Islanders' claim in the 1920s for secession from Finland was rejected because in the words of some League of Nations representatives, "To concede to minorities, either of language or religion, or to any fraction of a population the right of withdrawing from the community to which they belong, because it is their wish or their good pleasure, would be to destroy order and stability within States and to inaugurate anarchy in international life; it would be to uphold a theory incompatible with the very idea of the State as a territorial and political unity."[26]

Even though this conservative stance was known, during World Wars I and II and the interwar period, many secessionist movements existed in the British and French empires, the Arab world, Central America, the Philippines, and also appeared where they were least expected, namely, in France, Britain, Belgium, and Spain. World War II put a temporary cap on claims, as attention was focused on conflicts in Europe and East and Southeast Asia, but after the war the quest for self-determination became a dominant force in the Third World, as colonized peoples sought the removal of control by alien Europeans. Decolonization and independence was more or less met by the end of the 1960s, without much altering of the colonially derived boundaries of existing territories. More than eighty new States were created in the 1960s and early 1970s, generally with the blessing of the United Nations. Today all of the major former colonies have achieved independence, although many remain as nagging unresolved problems of decolonization (e.g., parts of Southwest, South, Southeast, and East Asia, the Pacific, and Africa, as well as in many European-settled States, where indigenous peoples were displaced—the United States, New Zealand, Canada, Australia, and Russia).

Demands continue for self-determination that would often involve secession from existing States, including some States in Africa and Asia that obtained independence recently. Under international law as generally understood today, however, self-determination can be granted but once to "a people" within any territory because self-determination has been restricted by two dominating, overriding principles—sovereignty and territorial integrity. Whereas it was legitimate for a whole "people" (defined by being located in territories delimited by European-imposed decrees) who were subjected to overseas colonial rule to seek self-determination—that is, by the application of the so-called salt water theory—it was not legitimate for

people who form a minority within a national territory to seek self-determination, whether on their own initiative or with help from any outside
power, for that would "dismember or impair" the existing State. In short,
international law remained grounded in the belief that the population of
the total national territory was paramount over parts of the population or
territory within it. Any departure from this, under international law as thus
understood, entailed the free choice of the majority of the total population
of the State out of whose territory the new State was to be carved, and not
just the sub-State minority who sought separation.

The rationale for the United Nations' stance and those of the Organization of African Unity and other international bodies is easy to understand.
These international bodies are made up of States, and the membership is
hesitant to permit parts of their respective territories to be hived off to create other States! To not speak out on this would be to invite or justify attacks on the territorial integrity of the existing member States. States thus
declared that self-determination could be applied only once to any territory,
that is, when independence from the colonial power was achieved. Exceptions to this (as in Bangladesh and, more recently, Eritrea) have been ex-
cepted by the international community who have claimed that the resulting
secessions were due to exceptional circumstances (i.e., excessive human
rights abuses in the first case and a remaining postcolonial element in the
second) and thus were not to be regarded as precedent setting.

Although the UN position reflects a desire for territorial stability, the formal application of the right to self-determination in all colonial territories
has had the potential to create absurd situations, as in some tiny island
"States." Self-determination has not always been "granted" to the peoples
within certain previously decolonized territories, such as in Goa (India) and
West Irian and East Timor (Indonesia). Any rights to self-determination by
the peoples of those territories were taken from them by the "invading"
States.

A different challenge to the territorial integrity of existing States occurred
recently outside a colonial setting when the USSR disintegrated and some
of its constituent territories declared independence. In addition, in former
USSR-dominated Eastern European States, as in Russia itself, governments
were overthrown—as expressions of internal self-determination—and the
States reconstituted. New States have been created out of territories that
were well defined within the former State, including Ukraine and Lithuania,
and from a splitting of territory, as with the recent creation of the Slovak
and Czech Republics out of Czechoslovakia, and the separation of Slovenia
from Yugoslavia. Most of these changes happened with remarkably little
violence, others with bitter conflicts, as in Azerbaijan and Armenia, where
there is as yet no clear territorial solution to a difficult situation. And,
worse, the atrocities in parts of former Yugoslavia stand as grim reminders

that the link between identity and territory can have disastrous consequences.

Identity

Self-determination necessarily involves issues of identity, for it is people, in
a particular territory, who express the desire for self-determination and,
possibly, secession and the creation of a new State, leading to the issue of
how the population is to be defined. There are many people who give primacy of belonging to some sub-State group, however defined and measured. Thus, for instance, in Canada some people within the province of
Quebec recognize their sub-State regional identity as having primacy, that
is, as a Quebecois identity. A population *subset* not having any allegiance
to an otherwise national identity is problematic in any State. The issue is
contentious all around the globe for there is a lack of homogeneity in almost all States, with perhaps Portugal and Iceland being the most clearly
defined exceptions, since they each have a fair fit between a homogeneous
population and the State territory. In that sense, they are perhaps the
only—or best—examples of the concept of "nation-State." Most States, in
contrast, have plural societies, with minorities.

The "minorities" or "national minorities" problem of the 1920s and
1930s was replaced after World War II with the demands by various sub-
State groups for recognition. "Ethnic" came into vogue in the 1950s and
especially 1960s to describe increasing numbers of sub-State identities de-
sirous of either some degree of self-government—not always apart from existing States—or, in the extreme, secession and the creation of new States.
The "ethnic kick" of the postwar years may have been in part a reaction
against Statewide modernization processes under way in many States,
which, it was believed, would help to create and strengthen *national* identities and override all other (sub-State) identities. Modernization did not necessarily bring about the expected areal fit between a nation and its State,
however, and even today a very weak sense of nation exists in most former
colonial territories. In the latter States (e.g., Sudan, Nigeria, India, and Sri
Lanka), certain sub-State allegiances remain very strong, and in many
States, civil (i.e., intra-State) wars have been and are being fought over who
has the ultimate right of control. Even in Western Europe sub-State region-
alisms are considerable and growing in vitality. Some sub-State groups demand devolution of power from the center (e.g., Scotland and Wales within
the U.K.); others call for secession and back their demands with violent acts
(e.g., in Euskadi within Spain). Interestingly, some demands for self-determination and secession are not based in an ethnic or national identity but
rather in a strong regionalism within parts of existing states. The incipient
independence movements in western Canada and Western Australia, fos-

tered by people of varied European heritage, who seem more driven by alienation from their respective States' geopolitical center than by their regional sense of "self," are examples.

The phrase "group territorial identity" has been suggested to encompass a whole variety of identities—regional, ethnic, tribal, and national.[27] The term is flexible and can be applied to any level in the hierarchy of attachments to territory, from small group, to a parochial localism, to a broader (but still sub-State) regionalism, to a nationalism (which may, of course, also be sub-State in areal focus), and even to an internationalism. The most potentially divisive level in this hierarchy of attachments is sub-State group politico-territorial identity, which, as a regionalism within the State, is an expression of self-determination. The use of the word politico here implies that the group's sense of identification has taken on a political dimension. A group politico-territorial identity can provide a threat to the State if its regionally based concerns become a sectionalism whereby political concerns are held to be more important than those that pertain to the whole State. If a sectionalism is potent and if accommodation cannot be reached between it and the existing State, then the sub-State group politico-territorial identity may seek and possibly achieve secession. If successful in achieving secession and if granted recognition by other States, then sovereignty could be said to have been achieved by means of an application of self-determination according to the wishes of the people involved. At that point the group politico-territorial identity would cease to be a regionalism, however, for it would have become a new territorial identity pertaining to a differently defined, smaller territorial frame, the new State. If international recognition were granted, it would be granted to the State, not to the people, for it is the State that would be accepted into the international system of States, not the territorially encompassed identity as such, or those who claim to speak on behalf of it.

Without doubt, a key group territorial identity in today's world is the nation. How is a nation defined? At one level it is simply a territorially based community of human beings who have like attributes, such as language and religion, with accepted societal structures, and a common (real or imagined) history, and so on. More provocatively, a British scholar has declared that "any territorial community, the members of which are conscious of themselves *as* members of a community, and wish to maintain the identity of their community, is a nation."[28] *Self* definition! But to permit all groups to self-define would, or could, create anarchy.

A we/they dichotomy is confronted with all group identities, especially in nations. Inherent in the definition of nation is the separation from other nations; the same applies to most other definitions. Inward or outward; security versus opportunity. This dichotomy, as an essential dilemma inherent in the concept of territory, also applies to a group's concept of "self." Consider how this may have an impact on a sub-State regionalism, which can be defined as a sub-State nationalism—as with (now historically) Ukrainians in the USSR. There is a danger—for the State—if inwardness can lead to an uncritical self-congratulatory self-glorification of the "nation," which in turn may lead to a blinkered, petty, narrow nationalism, with broader associations being ignored. In ignoring the value of broader associations the sub-State nationalism may determine that the only means for providing a secure place for the distinctive identity is through the creation of a separate sovereign State.

Clearly tension will exist between those who attach priority of belonging to the total community within an existing State structure and those who attach priority of belonging to a sub-State regional group that desires self-determination via secession. Those who do not identify with the broader existing State—like Quebecois separatists in Canada—claim that they want to live in a universe that conforms exactly to their group's narrow sense of "self." However, nationalists who seek to counter secession movements by sub-State regional groups (such as the Quebecois) believe that the challenge of coexisting obliges each community within a State that has a plural society to continually extend and surpass itself; in so doing, each group can then seek and achieve goals that, for each community taken separately, would never be attainable. Federalists in Canada, for instance, like nationalists in other federal and also unitary States, have a notion of oneness that pertains to the whole territorial unit, however poorly or well understood, and to all people who live within the State. The challenge for those who believe in "the nation," or at least in a national identity, is to get people within the State who have a different—minority—sense of self to accept the merit of unity with diversity. This challenge presents grave difficulties within numerous States in many parts of the world, where demands for separation and autonomy, or secession and independence, are considerable. But there are also instances where the challenge is not only being met, it is additionally being linked to new inter-State definitions of "self," as in Western Europe. Thus, interestingly, in the latter, there are people who are willing to extend their sense of self to a transnational entity, to a Europe-wide level of attachment even as their respective States fumble their way toward a united European Community. As they proceed, a new appreciation is developing for the limitations of sovereignty because the creation of a "new Europe," if it is fully formed, will lead to the granting of many functions to an extra-State administration that in the past would never have been entertained. However, there remain many Europeans who as, first, nationals (as Germans, Belgians, etc.) do not want to see the undermining of their States' sovereignty by handing over a critical mass of State functions and powers to EC bodies that have no clear legitimacy. The issue is perhaps toughest for Germans who, since unification of West and East Germany, have had to expand their sense of what it means to be German within the reconstructed State.

East Germany was relinked to West Germany following the fall of the Berlin Wall and the collapse of the communist regime. That event and related processes elsewhere in Eastern—now, again, Central—Europe and in the former USSR led to Lithuanians, Ukrainians, Slovenes, and others both claiming self-determination and receiving international recognition. At the same time, however, other peoples have been denied Statehood, as with a number of groups within Russia, even though they have a clear territorial base. The issue of double standards apply elsewhere too. Why, for example, were Kuwaitis and not East Timorese aided internationally when their territories were invaded, even though self-determination was clearly called for in the case of East Timor, and its "denial" was used as a pretext for retaliation in the case of Kuwait? It should be clear that when one deals with the political geography of self-determination, linked so clearly as it is to issues of identity and territory, important philosophical and moral issues must be faced. It can be concluded that we clearly live in exciting and yet difficult times, as old notions of separateness and of sovereignty are being challenged and debated.

Sovereignty

The previous section identifies changes in the concept of sovereignty, most notably the new extra-State linkages in Western Europe. To some politicians in certain States in Western Europe, the thought of "giving up" to a European Community decisionmaking that pertains to matters within a State (including such mundane concerns as the fat content of sausages and whether double-decker buses are acceptable) is tantamount to giving up sovereignty. Other politicians see the old rigid notions of sovereignty as being impediments to be overcome and are not threatened by a uniting Europe. Either way, people in Europe are actively dealing with new, freer concepts of sovereignty. Tension will continue between people who stress security at the "local" (now State) level versus those who now desire opportunity within an expanded European Community.

The concept of sovereignty is increasingly under attack from other sources too.[29] Therefore, it is being rethought due to such varied events and processes as the following:

- international human rights demands that cut against the notion that States are free to treat the people within their boundaries in ways the States' regimes think fit;

- the often invidious ways the capitalist world economy operates without due regard for the views and decisions of States' political leaders;
- a new international legal or moral persuasiveness about environmental matters that challenges States' rights to "develop" resources within their territories as they wish without consideration for what others think and expect—as with the decimation of "national" forests and, not so incidentally, indigenous peoples who inhabit some of those forests;
- bombing of a State's territory without a declaration of war;
- UN troop involvements within a State's territory, with or without the approval of the State's government;
- the declaration by a neighboring State or the UN that part of a State's territory is to be treated as a "no-fly" or "security" zone, despite the protests of the State in question;
- violent attacks against people and property within States' territories by terrorists who may be acting on behalf of, or with the support of, particular States' governments in other parts of the globe;
- a State claims extraterritorial rights or sanction activities in another State's territory.

Several of the points noted here involve acts and decisions that transgress States' sovereignty and territorial integrity. The latter, it should be remembered, form the core notions of Statehood and remain as the underlying bases for international society. However, the notion of a State's territory and its population being protected by virtue of sovereignty and territorial integrity is, in some regards, now increasingly passé due to such as the examples identified above. Yet, in so many ways, it still remains that sovereign governments make rules and regulations for people within their territories, so, not unexpectedly, there is tension. Indeed, the tension between "old" and "new" notions of sovereignty clearly promise that interesting times lie ahead.

Toward New Understanding

This chapter has touched on many concepts and processes that grow from a consideration of three essential characteristics of Statehood, namely, territory, population, and sovereignty. All three can be considered in terms of their legal character, but they become more interesting and challenging once they are approached from a more conceptual stance. These characteristics of Statehood thus can be seen to be dynamic and subject to change—undoubtedly because they always remain human constructs. They are not fixed, although they sometimes have been thought of as such, especially by those who have wished to use their substance to justify a particular State's actions! As the political geography of the late twentieth and early twenty-first centuries are considered, it is important to adopt a conceptual perspective so that understanding can be achieved concerning how the world's "one people" are organized in changing yet still divided ways.

Notes

1. B. C. Shafer, *Faces of Nationalism* (New York: Harcourt Brace Jovanovich, 1972).

2. The State (capital *S*) refers to a sovereign, independent, self-governed territorial organization; a state (lowercase *s*) is a sub-State territorial unit, such as Michigan within the United States, the latter being the State.

3. D. B. Knight, "Statehood: A Politico-Geographic and Legal Perspective," *GeoJournal* 28, no. 3 (1992): 311–318; M. I. Glassner, *Political Geography* (New York: Wiley, 1993).

4. L. Oppenheim, *International Law,* 8th ed., vol. 1 (London: Longman, 1955), p. 451.

5. R. D. Sack, *Conceptions of Space in Social Thought: A Geographic Perspective* (Minneapolis: University of Minnesota Press, 1980), p. 178.

6. See Glassner, *Political Geography,* note 3, pp. 61–71.

7. A. J. Day, ed., *Border and Territorial Disputes,* 2nd ed. (London: Longman, 1987).

8. A. F. Burghardt, "The Bases of Territorial Claims," *Geographical Review* 63 (1973): 225–245; A. B. Murphy, "Historical Justifications for Territorial Claims," *Annals of the Association of American Geographers* 80 (1990): 531–548; A. B. Murphy, "Territorial Ideology and International Conflict: The Legacy of Prior Political Formations," in *The Political Geography of Conflict and Peace,* ed. N. Kliot and S. Waterman (London: Belhaven, 1991), pp. 126–141.

9. T. M. Franck, "Clan and Superclan: Loyalty, Identity, and Community in Law and Practice," *American Journal of International Law* 90 (1996): 359–383.

10. J. Crawford, ed., *The Rights of Peoples* (Oxford: Clarendon Press, 1988).

11. D. B. Knight, "People and Territory or Territory and People: Thoughts on Post-Colonial Self-Determination," *International Political Science Review* 6 (1985): 249–250.

12. M. W. Mikesell, "The Myth of the Nation State," *Journal of Geography* 82 (1983): 257–260.

13. Y. Shain, *The Frontier of Loyalty: Political Exiles in the Age of the Nation-State* (Hanover: Wesleyan University Press, 1989).

14. A. Cassesse, *International Law in a Divided World* (Oxford: Clarendon Press, 1986), p. 78.

15. J. Anderson, "Nationalists, Ideology, and Territory," in *Nationalism, Self-Determination, and Political Geography,* ed. R. J. Johnston, D. B. Knight, E. Kofman (London: Croom Helm, 1988), pp. 18–39.

16. C. H. Williams and A. D. Smith, "The National Construction of Social Space," *Progress in Human Geography* 7 (1983): 502–518; A. B. Murphy, "Regions as Social Constructs," *Progress in Human Geography* 15 (1991): 22–35.

17. D. B. Knight, "Identity and Territory: Geographical Perspectives on Nationalism and Regionalism," *Annals of the Association of American Geographers* 72 (1982): 517.

18. V. Konrad, ed., "Nationalism in the Landscape of Canada and the United States," *Canadian Geographer* 30 (1986): 167–180; N. Johnson, "Cast in Stone: Monuments, Geography, and Nationalism," in *Political Geography: A Reader,* ed. J. Agnew (New York: Wiley, 1997), pp. 347–364.

19. W. Zelinsky, *Nation into State: The Shifting Symbolic Foundations of American Nationalism* (Chapel Hill: University of North Carolina Press, 1989); S. Cohen and N. Kliot, "Place-Names in Israel's Ideological Struggle over the Administered Territories," *Annals of the Association of American Geographers* 82 (1992): 681–695.

20. Y.-F. Tuan, *Space and Place* (Minneapolis: University of Minnesota Press, 1977), pp. 149–160; B. C. Lane, *Landscapes of the Sacred* (Mahwah, N.J.: Paulus, 1988); D. Lowenthal, "European and English Landscapes as National Symbols," in *Geography and National Identity,* ed. D. Hoosen (Oxford: Blackwell, 1994), pp. 15–38.

21. J. Gottmann, *The Significance of Territory* (Charlottesville: University Press of Virginia, 1973), p. 15, stress added.

22. See Knight, "Identity and Territory," note 16, p. 526.

23. See Shafer, *Faces of Nationalism,* note 1, p. 17.

24. R. J. Johnston, D. B. Knight, and E. Kofman, "Nationalism, Self-Determination, and Political Geography: An Introduction," in *Nationalism, Self-Determination, and Political Geography,* ed. R. J. Johnston, D. B. Knight, and E. Kofman (London: Croom Helm, 1988), p. 3.

25. See Knight, "Identity and Territory," note 16; M. W. Mikesell and A. B. Murphy, "A Framework for Comparative Study of Minority-Group Aspirations," *Annals of the Association of American Geographers* 81 (1991): 581–604.

26. See Knight, "Identity and Territory," note 16.

27. A. Cobban, *National Self-Determination* (Oxford: Oxford University Press, 1945), p. 48.

28. A. B. Murphy, "International Law and the Sovereign State: Challenges to the Status Quo," Chapter 12 in this volume.

29. Ibid.

Chapter Twelve

International Law and the Sovereign State System: Challenges to the Status Quo

ALEXANDER B. MURPHY

International law is a set of rules or principles that govern the actions and behavior of states. It is understood to encompass such matters as the right of one state to use force against another and the right of states to exercise control over ocean resources. But for all the importance of these kinds of international legal norms, a preoccupation with the role of international law as a simple regulator of state action can obscure a larger reality: that international law is the embodiment of widely accepted views of how territory should be organized and used. Indeed, the idea that the land surface of the earth should be divided up into more or less autonomous sovereign states is, itself, a principle of international law.

Comprehending the spatial organization of societies requires an understanding of the territorial ideas and arrangements expressed in and shaped by international law. How do particular ideas about the use of territory develop? How do they become implemented? And how does the implementation of those territorial concepts affect matters ranging from ethnic group relations to human alteration of the physical environment? This chapter focuses on the nature and significance of changing international legal norms with respect to territorial control. Particular attention is devoted to evolving concepts of state sovereignty and the ways in which international law reflects and shapes territorial organization and human-environment relations.

The State in Contemporary International Law

The roots of the contemporary system of international governance are generally traced to the fourteenth century, when Europe began to move out of an era in which territory was contested space over which feudal lords and kings vied for control. During this time the declining influence of the church, the rise of mercantilism, and the development of more sophisticated military technology allowed authoritarian rulers in some parts of Europe to claim and enforce relatively exclusive control over substantial domains.[1] In the succeeding centuries, European legal scholars began to elaborate principles to govern relations among these self-proclaimed independent territorial units. Inspired by precedents set in ancient Greece as well as the political realities of the time, scholars such as Francisco de Vitoria (1480–1546) in Spain and Hugo Grotius (1583–1645) in Holland argued for a system of international relations based on the absolute sovereignty of states. From the perspective of these founders of modern international law, any political authority who exercised effective control over a significant territory was entitled to govern that territory free of outside interference.

The principle of territorial sovereignty assumed wider formal status with the signing of the Peace of Westphalia in 1648. Each party to the treaties ending the Thirty Years' War agreed to honor the boundaries of the others and to refrain from interfering in their internal affairs. In so doing, a fundamental principle of the international system was established. Although this principle is often violated, it nonetheless continues to be the legal and intellectual foundation on which societies claim to base their international relations.[2]

Modern States

The commitment to state territorial sovereignty took on a new form with the spread of Enlightenment political ideas through Europe during the eighteenth and nineteenth centuries. Whereas the right to control territory had previously been viewed as the province of a ruling monarch, political legitimacy increasingly came to be seen as stemming from the rights of "the people." The people were understood to be a culturally cohesive community (a nation) that was entitled to control its own affairs. The Enlightenment worldview thus presupposed an international political order made up of discrete nations, each of which could be given its own autonomous territory, or nation-state. The Enlightenment ushered in an era in Europe during which sovereign nation-states were assumed to be the political geographic ideal. Nations were seen as distinct political and cultural communities with the right to control their own affairs in a territory that offered security and freedom from outside oppression. The notion of territorial sovereignty thus acquired a new kind of legitimacy, one premised on the ideological bedrock of "national" rights.

European Ideals

Europe embraced the nation-state ideal in the aftermath of the French Revolution. During the nineteenth and early twentieth centuries, the nation-state was incorporated into the national iconography of Europe's states, and great rhetorical deference was paid to it as Europe's great empires were carved up into states after World War I. Although there were many violations of national sovereignty during this period and although geopolitics played an extremely important role in the post–World War I negotiations over the fate of Central and Eastern Europe, states consistently acted in the name of facilitating a European order based on the principle of national territorial sovereignty. As such, the nation-states principle acquired the status of a fundamental norm of international relations in Europe.

Europe's global economic, political, and military reach meant that the European political order became the model for the emerging international state system. The Europeans did not treat their colonies as sovereign nation-states, considering them too "primitive" to have national communities and hence unable to enjoy the privileges of statehood. But the control that Europe and its North American offshoot exerted in international relations meant that any entity seeking freedom from colonial control and a place in the international order joined a system that, at least in theory, was made up of sovereign nation-states. Ironically, twentieth-century independence movements aimed at throwing off the yoke of European colonialism could only succeed if they claimed a status that itself was a European creation.

Questionable Sovereignty

The observance of state sovereignty has never been absolute, and more powerful states usually have been able to exert some control over the affairs of less powerful states. Indeed, the control exerted by the Soviet Union over states in Eastern Europe from the close of World War II to the late 1980s, the assistance provided by the United States to the Nicaraguan contras in the 1980s, and Israel's maintenance of a security zone in southern Lebanon all seem to belie the notion of a world order built on state territorial sovereignty. Similarly, neither Afghanistan during the 1980s nor Panama today possesses the same degree of sovereignty as that of the United States, Germany, or Japan.

Despite the many instances in which state sovereignty appears illusory, it remains a clearly articulated precept on which international relations are based. This can be seen in the major international legal instruments of the twentieth century. In 1919, the Covenant of the League of Nations bound all members "to respect and preserve as against external aggression the territorial integrity and existing political independence of all Members." Similarly, Article 2 (7) of the Charter of the United Nations, adopted after World War II, holds that "nothing contained in the present Charter shall authorize the United Nations to intervene in matters which are essentially within the domestic jurisdiction of any state or shall require the Members to submit such matters to settlement under the present Charter." Embedded in these and countless other international legal instruments is the notion that state territorial sovereignty cannot be abridged by international law.

Might Makes Right?

There exists, then, a conundrum. On the one hand, state territorial sovereignty appears to be a deeply rooted assumption of the international legal order. On the other hand, there are so many visible instances in which the sovereignty of one state is violated by another that it is tempting to dismiss that assumption as essentially meaningless. In analyses of international governance, this conundrum is often resolved by downplaying the issue of state sovereignty altogether: Territorial sovereignty is seen simply as a function of a state's economic and political might and is therefore unworthy of serious consideration.[3] Yet by adopting this approach, the ramifications of a historically rooted commitment to the principle of state territorial sovereignty are left unexamined. To make such a point is not to deny the variable character of territorial sovereignty or the growing challenges to state autonomy that have accompanied international law in the last few decades. Rather, it is to suggest that there has been no clean break with the historically rooted commitment to the ideal of state territorial sovereignty. International relations continue to be influenced by the assumption that the world is made up of largely autonomous nation-states. And that assumption is an integral part of the international legal order.

The Commitment to State Territorial Sovereignty

The broadest and most pervasive evidence that the commitment to state territorial sovereignty has some meaning is the fact that most issues and problems around the world tend to be conceptualized in state terms.[4] Despite pervasive evidence of the international nature of issues ranging from poverty to environmental degradation, the individual state is usually seen as the appropriate political geographic framework for confronting specific instances of these problems. Squatter settlements around Mexico City are thought to be Mexico's problem, the pollution of Lake Baikal Russia's problem, and the conflict between the Tamils and the Sinhalese Sri Lanka's problem. Indeed, for every instance in which international involvement occurs or is deemed appropriate, there are thousands in which the state is assumed to be the rightful controller of the situation. What this means in practice is that ecological, social, economic, and ethnic problems are generally confronted within political frameworks that do not bear any resemblance to the spatial-territorial dimensions of the problems themselves.

The norms of international diplomacy also confirm the influence that the concept of state territorial sovereignty has in the contemporary world. One remarkable feature of international relations is the general unwillingness of states to allow any party other than another state to sign an international agreement. This presumably reflects the assumption that states are the sole entities with the power or authority to assume international legal obligations. Even well-organized groups such as the Palestine Liberation Organization have generally been able to participate in negotiations over international agreements only as members of another state's delegation. Moreover, international bodies have often refused to recognize claims that do not arise from states. Despite the avowed commitment of the United Nations to national self-determination, it has refused to support the claims for territorial autonomy of such groups as the Kurds and the Biafrans since the territories they seek to control lie within the boundaries of existing states.

Assumptions about the preeminent role of states in international relations are also evident in the foreign policy practices of most states. Typically, such policies are much more likely to be driven by reactions to the positions of other state governments, rather than to nongovernmental actors within those states. Thus, during the 1970s and 1980s, U.S. foreign policy toward Iran was far more influenced by actions of the shah and then the Ayatollah Khomeini than by anything else that was going on in that country. As a result, diplomats focus on the workings of government in foreign capitals, while often ignoring regional issues and problems.

The extent to which notions of state territorial sovereignty govern international relations is revealed in the skepticism that is often expressed concerning the idea of international law itself. How can there be international law, many ask, when the only real power rests with states? This view improperly equates law with effective, centralized enforcement—although such an approach would disqualify many domestic laws from being regarded as true laws—and it ignores the role of law as a standard against which most parties measure right and wrong. Moreover, it fails to take into consideration the pains to which states go, in most instances, to comply with international legal norms as well as the repercussions that can fall on a state for lack of compliance ranging from economic sanctions to military intervention.

The strength of the myth that there is no true international law reflects the continuation of the centuries-old doctrine that state political authorities should be the arbiters of what happens within their boundaries. This myth, in turn, greatly complicates the task of garnering support in many countries for international initiatives. Indeed, it is one of the great impediments to the adoption of more sweeping international agreements over the world's oceans and seas.

Territorial Conflicts

The power that the concept of state sovereignty holds in the modern world can be strikingly demonstrated in the role it has played in international territorial conflict. Despite the invocations against attacking the territory of neighboring states found in international legal agreements, more than half of the world's states have been involved in some sort of territorial dispute with a neighbor since the close of World War II.[5] Such disputes may seem to illustrate the meaninglessness of the principle of state territorial sovereignty, but a more careful examination of them actually reveals the important role that the sovereignty principle plays in shaping the location and character of international conflicts.

During the past forty-five years, interstate territorial conflicts have almost always involved territory that one state could claim to have been wrongfully taken from it at some prior time.[6] In some cases, the claim may be weak, but usually there is some historical period when the territory was either within its domain or within the domain of a political-territorial antecedent of that state (a colony, an administrative territory within an empire). Why should this be? If interstate relations are based solely on political and economic power, why are the territories in dispute not simply those that offer the greatest riches or strategic advantages to states? Valuable territories often are in dispute, of course, but why has Japan pursued so aggressively its claim to a few small, sparsely inhabited islands northeast of Hokkaido (the Northern Territories) instead of more economically and militarily valuable islands farther north? And why does Venezuela persist in laying claim to agriculturally unproductive rain forest areas in Guyana instead of the oil-rich area of northern Colombia just across the border? The answers lie in the continuing vitality of state territorial sovereignty as a principle of international relations.

The pursuit by one state of a claim to territory in another requires that some sort of explicit justification be advanced, to rally support for the cause and to avoid international isolation or condemnation. To be successful, the claim must be "fair." Notions of fairness, however, are subjective and constantly changing. Since World War II, the dominant view has been that no state has the right to seize the territory of another. This principle, incorporated in most major international agreements, is a direct reflection of a commitment to territorial sovereignty.

The only generally recognized exception to this rule is that a state whose territory has been wrongfully seized by another may act to retake the "stolen" territory. Thus, if state X marches into state Y and seizes a third of its territory, state Y is generally thought to be justified in mounting an action to retake the territory, even if it takes some time for the necessary forces to be assembled. Without a statute of limitations for territorial "theft," a restitution claim can always be made without directly challenging dominant international understandings of justice.

Justifying Claims

A variety of economic, political, cultural, and strategic motives are behind most interstate conflicts over territory.[7] Yet with few exceptions, the only stated reason for pursuing a territorial claim is to regain wrongfully appropriated land. Government leaders rarely declare, either before their own people or in front of the world community, that they are entitled to territory in a neighboring state because it would expand their domestic oil reserves or allow them to exert more effective control over the surrounding seas. Rather, whether it is Argentina claiming the Falkland Islands, China claiming territory across the Amur River in Russia, or Togo claiming part of eastern Ghana, leaders argue that they are merely seeking to retake land that historically belonged to their state. Only by raising this type of argument can the state hope to gain national and international support for its cause, since other possible arguments would overtly challenge notions of justice rooted in the ideology of state territorial sovereignty.

Articulated justifications for territorial claims, even if little more than hollow rhetoric, are not just meaningless statements; they have a significant impact on the pursuit of such claims and have shaped the geography of interstate territorial conflict itself. And because historical arguments are normally needed to justify territorial claims, states whose boundaries have not undergone significant changes are unlikely to raise extrastate claims to territory. This has been the case throughout much of sub-Saharan Africa, which has experienced considerably fewer interstate conflicts over territory in the post–World War II era than has the Middle East, Latin America, or Asia.[8]

On a smaller scale, the reliance on territorial justifications can affect the nature and extent of territories in dispute. Thus, Ecuador's claim to northern Peru (see Figure 12.1) cannot be understood merely as a quest to control an oil-producing region, as some have claimed. If Ecuador were con-

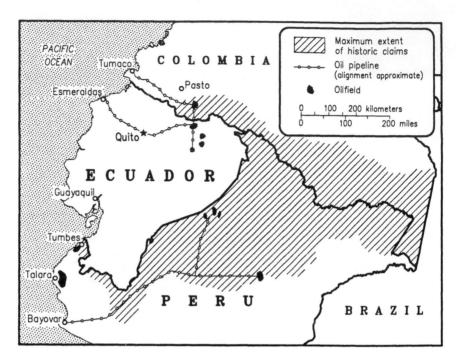

FIGURE 12.1   *Territory in dispute between Ecuador and Peru.*

SOURCE: Alexander B. Murphy, "Historical Justifications for Territorial Claims," *Annals of the Association of American Geographers* 80, no. 4 (1990): 538.

cerned merely with enhancing its oil supplies, it could claim the oil fields around Talara in Peru or those near Pasto in Colombia. The sole claim that Ecuador can hope to sustain in accordance with modern conceptions of justice is to the territory that was controlled by its colonial antecedent for a time during the Spanish occupation and that was awarded to Peru under questionable circumstances in a 1942 protocol witnessed by four other guarantor states. Since the legitimacy of Ecuador's claim would be compromised if its actions were not consistent with its claim, the state can push for no more and no less than the area that was in dispute at the time of the 1942 protocol.

The influence of the doctrine of state territorial sovereignty arguably extends beyond interstate conflicts over territory to the willingness of states to intervene in distant conflicts. A remarkable feature of the U.S.-led effort to retake Kuwait after Iraq's invasion of that country in 1990 was the large number of countries involved. Many have argued that the U.S. desire to retake Kuwait was motivated mostly by strategic and economic concerns. This may be true, but it does not explain why the United States was able to garner such widespread support from other governments. Nor does it explain why Washington has not sent troops into other places of great economic and strategic significance, such as Egypt after the Suez Canal was closed in 1967 or Yugoslavia after the outbreak of civil war. The answer to these puzzles almost certainly lies in the powerful role that the concept of state sovereignty continues to play in the world today.

When Iraq invaded and annexed Kuwait, it temporarily obliterated from the map a recognized sovereign state. Although Iraq sought to advance a weak historical argument after the fact, the invasion was never widely regarded as an attempt to retake wrongfully appropriated territory. Rather, it was seen as a direct and complete violation of both Kuwait's sovereignty and one of the core principles of international law. This argument provided the primary pretext for justifying military involvement by the United States, and it was essential to orchestrating widespread support from other countries.

The concept of state territorial sovereignty thus plays an important role in influencing the ways in which international problems are understood as well as the foreign policy practices of states and the norms that govern international behavior. Consequently, it is misleading to dismiss the ideal of state sovereignty as a total sham simply because it is violated with some frequency. Instead, we must recognize the ideal for what it is: a historically rooted concept of human territorial organization that continues to influence, albeit sometimes in indirect ways, international actions and behavior. Although state sovereignty is still "sacred" in some respects, important developments in international governance have unfolded in recent decades that impinge on significant aspects of it.

Growing Challenges to State Sovereignty

As the interdependencies that characterize our world increase in complexity and visibility, more and more pressure is being brought to bear on sovereignty as an underlying precept of the international order.[9] In fact, it is difficult to think of any significant social problem that does not have some sort of international dimension, be it environmental pollution, civil strife, human rights violations, or government debt. Moreover, supranational aspects of these problems are fueling unprecedented growth in international law. This growth, in turn, is precipitating a subtle shift away from the state as the spatial unit within which problems are assumed to be most appropriately confronted. At the same time, the rise of substate nationalism and regionalism in the post–World War II era is challenging the sanctity of the state. Although traditional understandings of state sovereignty continue to color international relations in fundamental ways, it is an open question whether a shift away from the state signals the beginning of a more fundamental change in the spatial-territorial assumptions that undergird the international system.

Three important international legal developments exemplify the changing position of the state in international relations: the elaboration of a comprehensive set of rules governing ocean use and management, the incorporation of human rights principles into international law, and the rise of the European Community (now European Union) as an international actor. In each case, traditional notions of state sovereignty have been challenged, with implications for the current spatial-territorial order.

Control and Use of the Seas

Historically the open seas have not been subject to state jurisdiction. Resources of the open seas were assumed to be inexhaustible, and given their immensity, conflicts over use were minimal. Through the 1950s, states generally exercised exclusive control over the so-called territorial waters within 3 nautical miles of their coastlines.[10]

The traditional treatment of the open ocean as a "commons" did not grow out of an abrogation of the principle of state sovereignty. It merely reflected the lack of any perceived need to exercise control over such a vast and limitless resource. Conflicts developed over various enclosed or semi-enclosed seas, and some states sought to extend the limits of their territorial waters, but the open seas were of little immediate concern to most states. The situation began to change in the twentieth century, however, with the rapid expansion in oceangoing traffic, the discovery of valuable resources in the continental shelves, the overexploitation of certain ocean resources, and the growth of such potentially damaging activities as offshore oil drilling and the dumping of waste materials.

The initial reaction to these developments confirmed the traditional role that state sovereignty played. In 1945, following more than two decades of discussion about the need to codify the rights of individual states to ocean resources, the United States announced that it intended to exercise exclusive control over the resources of the continental shelf and adjacent waters within 200 miles of its coastline.[11] The United States made clear that it did not intend to limit navigation outside of its traditional territorial waters, but its actions prompted a spate of similar claims by other states. Some states did not stop with assertions of control over resources; in 1947 Chile and Peru claimed exclusive sovereignty over the 200-mile-wide stretch of ocean extending from their shores.

Despite the adoption of conventions dealing with certain aspects of international control and use of the oceans over the succeeding two decades, efforts to gain widespread support for a single, comprehensive approach to ocean governance were unsuccessful. By the late 1960s, states were making increasingly extravagant claims to ocean resources, and there was growing public concern over pollution, overfishing, and military use of the seas. In 1967, the United Nations established a committee to explore the peaceful uses of the ocean, and in 1970, the UN General Assembly adopted a resolution calling for the seabed and ocean floor to be treated as a "common heritage of mankind." These events set the stage for a series of important international meetings on the law of the seas between 1973 and 1982, organized under the auspices of the United Nations. The meetings culminated with the adoption, by an overwhelming majority of participants, of the United Nations Convention on the Law of the Sea.

In brief, the convention calls for a uniform 12-mile territorial sea over which states have exclusive authority (with the exception of "innocent passage") and a 200-mile exclusive economic zone (EEZ), in which states can exercise sovereign rights over resources. The remaining seas are treated as commons, but not in the sense that they are open to unrestricted use by any state. Rather, they are to be used for peaceful purposes and are subject to rules on resource exploitation and scientific research established by the International Seabed Authority. The convention represented, at least in part, a departure from the traditional idea that individual states have ultimate control over the surface of the earth. Instead of dividing up the oceans into national territories, the convention calls for the establishment of a cooperative system with rules articulated not by the individual states but by an international authority.

Many issues were left unresolved and much in the convention reaffirms the principle of state sovereignty—including the nature and extent of the territorial waters and the EEZ. Even though the overwhelming majority of states signed the convention, several influential states—notably the United States, the United Kingdom, and West Germany (now Germany)—refused to do so. The United States in particular has objected to the convention's restrictions on deep seabed mining. Yet despite these problems, the convention represents a significant milestone in the management and use of the oceans. Even nonsignators such as the United States accept that much of the convention embodies customary international law, and to that extent, they have been willing to abide by it. In 1983 the United States even declared its own EEZ in accordance with the principles set forth in the convention.

From a broader perspective, the convention demonstrates a willingness on the part of the international community to confront one major issue in a manner that transcends traditional concepts of state sovereignty. It provides the frame of reference against which decisions and actions with respect to the oceans are now judged. No longer are the open seas seen as domains of exclusive state sovereignty or as realms within which states can do what they please. Instead, increasing attention has been directed to such issues as the functioning of international regulatory bodies, the need for regional co-operation, and the future uses of the oceans.

Human Rights

States have long recognized an international legal obligation to respect the basic rights of visitors from foreign countries. But until quite recently, it was assumed that the treatment of nationals within a state was beyond the purview of international law. Indeed, there is no better example of the status of state territorial sovereignty than the free hand most states traditionally had in dealing with their own citizens. The first real challenge to this did not come until World War II, when "the atrocities of the Third Reich brought home that large-scale deprivations of human rights not only decimate individuals or groups but endanger peace and security."[12] As a direct consequence, the protection of human rights became one of the primary stated objectives of the United Nations. An International Commission of Human Rights was convened after the war, and in 1948 the UN adopted without dissent the commission's Universal Declaration of Human Rights (several countries did, however, abstain, including the Soviet Union, Yugoslavia, and Saudi Arabia).

The declaration set forth fundamental principles for the treatment of peoples in all states, and although not legally binding, it has been quite influential. Over the past four decades, many countries have adopted statutory and constitutional provisions consistent with it. The declaration also led to the adoption of several important international conventions on human rights in the 1960s and 1970s, and current domestic and international legal instruments pertaining to human rights demonstrate a widespread commitment, at least in theory, to the idea that states have a duty to safeguard the basic human rights of their citizens. States are believed to be obligated to protect their citizens, to guarantee them equal protection of the law, and to grant them fundamental political, social, and economic rights.[13] Willful violations of these obligations may provide sufficient grounds for international condemnation and even sanctions.

The precise international legal status of much of human rights law is disputed. In practice, the commitment to state sovereignty is still sufficiently strong that human rights violations by themselves rarely provide adequate grounds for international intervention. Nonetheless, international recognition of human rights principles has fundamentally challenged the nineteenth-century view that states have unlimited freedom to treat their citizens in whatever manner they please. Some states even incorporated international understandings of human rights into their domestic legal codes.

The imposition of economic sanctions against South Africa in the 1980s largely grew out of international reaction to the state supported system of apartheid. Crude and undiplomatic actions have also been taken in response to reports by governmental and nongovernmental organizations on the human rights records of particular governments. One of the stated reasons for deferring Turkey's application to join the European Community, for example, was its weak human rights record.

The extent to which an international commitment to human rights has undermined traditional notions of sovereignty is most clearly seen in the international response to the plight of the Kurds and the Shiite Muslims in Iraq after the Gulf War. The plight of the Kurds in northern Iraq prompted the United Nations Security Council to adopt a resolution in April 1991 stating that a government could be required to accept foreign aid. This, in turn, was used as the legal basis for the U.S., British, and French establishment of a "security zone" in northern Iraq for the returning Kurds. In a further effort to protect the rights of minority groups in Iraq, internationally sanctioned "no fly" zones were established first in northern Iraq where the Kurds live and then in the Shiite-dominated south. Any Iraqi military plane that flies in these areas is threatened with attack from outside forces. Since these measures followed a major armed conflict, they are, in a sense extraordinary. But they also signal that there are limits to international tolerance of what a recognized state can do against its citizens.

State sovereignty has not, however, become irrelevant in the human rights realm; most states are still reluctant to become involved in human rights issues within other states without some kind of external or special provocation. Furthermore, human rights principles are now enshrined in the international state system: "as a corollary of its membership in the international community, every state is under a duty to respect the human rights and fundamental freedoms of every human being and to subject itself to legitimate measures of international scrutiny that the international community is entitled to utilize to ensure protection of human rights and fundamental freedoms."[14] Human rights advocacy may not be universally accepted, but it is indicative of a trend toward viewing the protection of such rights as an obligation that transcends the sovereign territorial rights of states.

The Special Case of the European Union

Most efforts at encouraging regional cooperation have had little impact on state sovereignty. The European Union (EU), formerly the European Community (EC), is a marked exception. From a modest beginning in the late 1950s, the EU has come to represent by far the most ambitious effort at interstate regional cooperation undertaken in modern history. The fifteen member states (the Netherlands, Belgium, Luxembourg, France, Germany, Italy, Denmark, the United Kingdom, Ireland, Greece, Spain, Portugal,

Sweden, Finland, and Austria) have vested substantial authority in the central institutions of the European Union on matters ranging from agricultural subsidies to environmental pollution.

The EU is often thought of simply as a common market—an interstate area in which the marketing of goods and services is unaffected by common international boundaries. In fact, a common market was the major thrust of European integration in its early stages, but it soon became clear that closely coordinated economic policies among the member states were required. Such matters as environmental quality standards and regional socioeconomic differences also brought the integration process into sharper focus. As a result, the EU gradually developed more powerful central governmental institutions with authority to promulgate rules and regulations.

EU-wide laws now deal with the movement of goods and people across boundaries, monetary exchange rates, basic environmental standards, transportation, working conditions, regional development, and much more. Despite the broad reach of EU legislation, the majority of the rules do not directly challenge the sovereignty of its member states. Until very recently, most major EU decisions could not be made without the unanimous support of the member countries, and significant realms of economic and political decision making have remained within their purview. To conclude that the EU has few implications for the doctrine of state sovereignty, however, is to ignore some fundamental institutional developments within Europe over the past two decades.

When the central institutions of the EU were established, substantial authority was vested in a bureaucracy located in Brussels. This bureaucracy, which encompasses some 14,000 functionaries who do not act officially on behalf of individual countries, comprises the staff of the Commission of the European Communities. Although lacking the power to adopt legislation in many areas, the commission is the driving force of the EU because it alone makes proposals for rules and regulations. It also is responsible for implementing EU rules and managing the EU's budget.

The commission's proposals can only be adopted by the Council of Ministers, made up of representatives from member states. Traditionally, any proposal involving an essential interest of a member state could only be adopted by a unanimous vote. Although the essential interest doctrine was frequently invoked, many important decisions during the 1970s and 1980s were made simply by majority vote. But with the adoption of the Single European Act, the requirement for unanimous consent has been greatly curtailed. Now, all that is necessary to pass major legislation in such areas as farm subsidies and environmental regulation is a qualified majority (whereby the votes of larger countries are weighted more heavily than those of the smaller countries, allowing two or more larger countries to block an affirmative vote).

In practice, EU member states have ceded important aspects of their sovereignty to these centralized institutions. Proposals considered for implementation, for example, emanate from an extrastate bureaucracy, and many are enacted without unanimous consent; nonetheless, they are accepted as law by the member states. Moreover, the decision to pursue monetary union within much of the EU signals an even greater cession of traditional sovereign economic powers by states, and the adoption of the 1997 Treaty of Amsterdam moves the union closer to a common foreign and security policy. The hesitancy of countries such as the United Kingdom to join the monetary union shows both the continued strength of national differences within Europe and the concerns that exist about strengthening the Brussels bureaucracy. But any significant revocation of the core economic and social powers that have been vested in the institutions of the EU is unlikely.

The implications of European integration for state sovereignty extend beyond the movement of the European Community into realms traditionally controlled by states. As barriers between member states continue to weaken, the ability of local governments and businesses to forge links across international boundaries is enhanced. Cross-border regional cooperation schemes are growing, such as those along the Upper Rhine between France and Germany, and significant economic and cultural links are developing among geographically dispersed regions within different states.[15] A striking example of this is the cooperation agreement entered into by Rhône-Alps (France), Baden-Württemberg (Germany), Catalonia (Spain), and Lombardy (Italy) in 1990 (Figure 12.2). Initiatives along these lines expand the range of international activity that is beyond state government control, as do EU programs that encourage cross-border and interregional cooperation. Although most cooperation schemes still operate with significant state involvement, the importance of state sovereignty is likely to diminish as those schemes expand.

Since the EU itself is a creation of international law, and since it, in turn, makes international laws, EU developments have implications beyond Europe. One important indication of this is that the EU has increasingly been treated by other states as an international actor analogous to a sovereign state. During the crisis in Yugoslavia in the early 1990s, it was the EU, as much or more than any individual state, that was looked to as a potential intervener. Precedents established by the EU are also regarded in other parts of the world as bases for interstate cooperation. The Commonwealth of Independent States (most of the republics of the former Soviet Union) as well as the free trade agreements between Mexico, Canada, and the United States are institutional structures derived, at least in part, from the EU model. The future stability of these regional cooperation schemes may be in doubt, but the institutional structure of the EU will still be viewed as a pioneering, forward-looking response to regional issues and problems. Since

FIGURE 12.2   Regions participating in the "Four Motors" Agreement.

SOURCE: Alexander B. Murphy, "Emerging Regional Linkages Within the European Community: Challenging the Dominance of the State," *Tijdschrift voor Economische en Sociale Geografie* 84, no. 2 (1993): 112.

the EU embodies a departure from the traditional norms of international governance, its acceptance represents a formidable challenge to the notion that a successful international state system must be one premised on state territorial sovereignty.

Conclusion

Despite indications that its ideological power is weakening, state sovereignty remains an important organizing force in the modern world. The preeminent position it holds is revealed most strikingly in the place that the political map occupies in our conceptualizations of the world around us. If people are familiar with any map of the world at all, it is likely to be the map of so-called sovereign states. How many people, when they think of South America, have a mental picture of a continent divided up into different physiographic regions or different ethnic areas or different vegetation zones? All these are interesting and important divisions within South America, yet most people who know anything at all about the continent think of a map showing Brazil, Argentina, Chile, and the like. Moreover, our descriptions of the world are based on the political map when references are made to the Ganges Plain or the humid subtropical climate zone of South Asia, the states of India, Pakistan, and Bangladesh immediately come to mind.

The habitual use of political maps as frameworks for thinking about the world reflects a tacit assumption: The units shown on those maps are meaningful spatial compartments for considering most international issues. This assumption, which has its roots in the concept of state sovereignty, has been woven into the norms of international governance over the past few centuries and is likely to be with us for some time to come. But human society and its norms of governance are not static. As the twenty-first century looms, the traditional concept of the state is being challenged by several key trends: growing economic interdependencies among states; the development of transportation, communication, and information technologies that facilitate international linkages; the rise of substate nationalism; and the growth of an extrastate corporate culture. With those challenges becoming more and more apparent, the conceptual hegemony of state sovereignty is being undermined. Although the legacy of the Peace of Westphalia remains very much with us, its future is increasingly clouded.

Notes

1. Jean Gottmann, *The Significance of Territory* (Charlottesville: University of Virginia Press, 1973).

2. See generally Thomas J. Biersteker and Cynthia Weber, eds., *State Sovereignty as Social Construct*, Cambridge Studies in International Relations (Cambridge: Cambridge University Press, 1996).

3. Peter J. Taylor, "Contra Political Geography," *Tijdschrift voor Economische en Sociale Geografie* 84, no. 2 (1993): 82–90.

4. John Agnew, "The Territorial Trap: The Geographical Assumptions of International Relations Theory," *Review of International Political Economy* 1, no. 1 (1994): 53–80.

5. Alan J. Day, ed., *Border and Territorial Disputes,* 2nd ed. (Harlow, England: Longman Group, 1987).

6. Alexander B. Murphy, "Historical Justifications for Territorial Claims," *Annals of the Association of American Geographers* 80, no. 4 (1990): 531–548.

7. Andrew F. Burghardt, "The Bases of Territorial Claims," *Geographical Review* 63, no. 2 (1973): 225–245.

8. Herman Van der Wusten, "The Geography of Conflict Since 1945," in *The Geography of Peace and War,* ed. D. Pepper and A. Jenkins (Oxford: Basil Blackwell, 1985), pp. 13–18.

9. Stephen D. Krasner, "Sovereignty: An Institutional Perspective," *Comparative Political Studies* 21, no. 1 (1988): 66–94.

10. Lewis M. Alexander, "Geography and the Law of the Sea," *Annals of the Association of American Geographers* 58, no. 1 (1968): 177–197.

11. Martin I. Glassner, *Political Geography,* 2nd ed. (New York: John Wiley & Sons, 1996).

12. Lung-Chu Chen, *An Introduction to Contemporary International Law: A Policy-Oriented Approach* (New Haven: Yale University Press, 1989), p. 204.

13. Ibid.

14. B. G. Ramcharan, "Strategies for the International Protection of Human Rights in the 1990s," *Human Rights Quarterly* 13 (1991): 1655–1669.

15. Alexander B. Murphy, "Emerging Regional Linkages Within the European Community: Challenging the Dominance of the State," *Tijdschrift voor Economische en Sociale Geografie* 84, no. 2 (1993): 103–118.

References

Alexander, Lewis M. "Geography and the Law of the Sea." *Annals of the Association of American Geographers* 58, no. 1 (1968): 177–197.

Burghardt, Andrew F. "The Bases of Territorial Claims." *Geographical Review* 63, no. 2 (1973): 225–245.

Chen, Lung-Chu. *An Introduction to Contemporary International Law: A Policy Oriented Approach.* New Haven: Yale University Press, 1989.

Day, Alan J., ed. *Border and Territorial Disputes,* 2nd ed. Harlow, England: Longman Group, 1987.

Glassner, Martin I., and Harm J. de Blij. *Systematic Political Geography,* 4th ed. New York: John Wiley & Sons, 1989.

Karatochwil, Friedrich, Paul Rohrlich, and Harpeet Mahajan. *Peace and Disputed Sovereignty: Reflections on Conflict over Territory.* Lanham, Maryland: University Press of America, 1985.

Krasner, Stephen D. "Sovereignty: An Institutional Perspective." *Comparative Political Studies* 21, no. 1 (1988): 66–94.

Murphy, Alexander B. 1990. "Historical Justifications for Territorial Claims." *Annals of the Association of American Geographers* 80, no. 4 (1990): 531–548.

_____. 1993. "Emerging Regional Linkages Within the European Community: Challenging the Dominance of the State." *Tijdschrift voor Economische en Sociale Geografie* 84, no. 2 (1993): 103–118.

Ramcharan, B. G. "Strategies for the International Protection of Human Rights in the 1990s." *Human Rights Quarterly* 13 (1991): 1655–1669.

Taylor, Peter J. In press. "Contra political geography." *Tijdschrift voor Economische en Sociale Geografie.*

Van der Wusten, Herman. "The Geography of Conflict Since 1945." In *The Geography of Peace and War,* ed. D. Pepper and A. Jenkins, pp. 13–28. Oxford: Basil Blackwell, 1985.

Williams, Allan M. *The European Community: The Contradictions of Integration.* Oxford: Basil Blackwell, 1991.

Chapter Thirteen

Global Ecopolitics

PHYLLIS MOFSON

The geopolitics of international environmental decisionmaking are rapidly changing as the implications of environment-related problems exceed local and national concerns. The broad range of actors—from governments to activists to the media—are taking on enhanced and more diverse roles, traditional alliances are breaking down, and new partnerships are being forged. Environmental problems are occupying higher priority positions on government agendas worldwide; they have focused public attention on assessing responsibility for pollution and allocating cleanup costs. The complexity of transnational environmental problems is even changing the role of science in international policymaking.

Diplomats are now working together in international fora to solve common environmental problems and protect common resources. They are being forced to recognize and to move beyond a long list of their differences that have traditionally hindered environmental cooperation, including cultural perspectives, economic priorities, and domestic political agendas. In addition, governments have different roles in polluting and regulating nongovernmental polluters and differ in their commitments to international cooperation. These vast differences make international agreements for implementable environmental solutions a much greater challenge than they are at the national or local levels.

The creation, impacts, and solutions of a given environmental problem change as the problem involves increasingly larger and more complex ecological and political systems. For example, the widespread use of high-sulfur brown, or lignite, coal for energy in Eastern Europe has created large strip mines and, in the areas surrounding coal-burning power plants, some of the world's most polluted air. On a regional level, the burning of brown coal contributes to acid rain that damages Europe's forests. The global im-

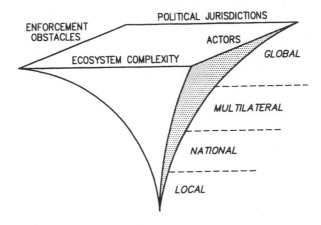

FIGURE 13.1   Ecopolitical hierarchy.

SOURCE: Office of the Geographer, U.S. Department of State.

pact of many inefficient brown coal–burning plants has been to exacerbate the buildup of greenhouse gases in the atmosphere and contribute to accelerated global climate change. Just as an environmental problem, such as dependency on brown coal, changes as it shifts from the local to the international levels, diverse political systems must also be reconciled and coordinated before workable solutions can be found and implemented.

Ecopolitical Hierarchy

A model of environmental politics can be constructed as an upside-down pyramid with four layers representing local, national, international/multilateral, and global ecopolitical scales or stages (see Figure 13.1).[1] Each of the pyramid's four sides represents an attribute of a given environmental problem: the complexity of the ecological systems involved, the number of institutional actors, the number of political jurisdictions, and institutional obstacles to forging and implementing solutions. As a problem moves from one level to the next, each of these variables becomes more numerous and their interactions more complex. Furthermore, perceived impacts and solutions at one level may contradict those at the next level.

Consequently, as a particular problem moves up the ecopolitical hierarchy, resistance to political action also becomes more intransigent. The jump from the national to the international level may be the most challenging of all. Some global issues—for instance climate change—involve most national governments and nongovernmental organizations, in addition to several supranational agencies.

Local Level

The local level, represented by the first layer in the inverted pyramid, embodies its own hierarchy of ecological and political relationships. These move from single neighborhoods up through such first-order administrative units as states and provinces. As the subnational unit increases in size, it inherits the conflicts among the smaller units that compose it. Despite these complexities, the cause and effect relationships that pertain to local environmental problems are relatively clear, and the roles of the actors relatively well defined.

National Level

At the national level, these relationships become slightly more ambiguous. At this level, environmental politics involve the actors from subnational units as well as federal governments, national nongovernmental organizations (NGOs), and large industries. Formulating national policy is particularly complicated when the source and impacts of a given environmental problem are not equally shared among the subnational units; a good example of this is the acid rain problem in the United States.

The role of the national government in pollution and cleanup programs is determined largely by the institutional and ideological structures that define the state. The major environmental role of national governments in most industrialized, capitalist countries is to regulate private industry. Although governments do pollute (such as defense and energy industry sites), much pollution in these countries is a product of privately owned economic activities. Environmental damage is the result of a variety of inputs, from the byproducts of industrial production to high consumption and accompanying waste generation from residential populations.

In centralized governments, by contrast, most major industries are controlled by ministries, making the state the primary polluter. Since the state is also the regulator of polluting industries, this situation is analogous to the fox guarding the henhouse. In some cases, such as China, the role of the state as polluter and regulator is complicated by the fact that agricultural and other small—often household-level—economic units are major sources of pollution. The ability of a national government to influence individual behavior is limited by often-conflicting goals; the goal of reducing agriculture-related pollution, for instance, may conflict with governmental demands to increase agricultural production.

Most environmental problems in the developing world can be traced to rapidly growing, impoverished populations. National governments of poor, agriculture-based economies frequently have less formalized control over subnational units and individuals than in the industrialized world. They also lack financial and other resources to implement the sometimes rudimentary environmental regulations that may have been legislated. Even when governments do have the means to regulate polluting activities, they often lack the will to do so. When environmental protection is perceived as detracting from economic development plans, it is often assigned a low priority on political agendas.

The formerly centralized governments of the countries of Eastern Europe and the former Soviet Union represent a unique combination of inherited and recently acquired characteristics. They carry the legacy of centralized government programs that emphasized large-scale industrial and security-related projects over environmental and other quality-of-life concerns. Data released after the dissolution of the Soviet Union indicate, for example, that industrial air pollution in 103 former Soviet cities has left them with more than five times the acceptable levels of pollutants.[2] The people of Eastern Europe and the former Soviet Union also continue to be subjected to some of the world's most severe water pollution, acid rain damage, and exposure to radiation.

Although these countries have inherited pollution problems that are characteristic of their former totalitarian (state-as-polluter) government structures, their current economies (like many Third World countries) are plagued by a lack of financial resources and by competing environmental and developmental priorities. Emerging capitalism in the former East Germany and the former Soviet Union means that several former-communist countries are developing a new set of polluters in the private industrial sector, and their governments are beginning to grapple with Western-style problems of regulating nongovernment industry while promoting the growth of the private sector.

International Level

Forging environmental agreements at the international level is complicated by the variety of roles and characters of national governments, as well as by the subnational layers of actors and roles below them. Cooperation at this level is made even more complex by almost universal resistance to surrendering any degree of national sovereignty to a supranational authority. International environmental action is further constrained by its high price tag; disagreement over who should pay to rectify damage already done, and to prevent further degradation, can paralyze negotiations.

Global Level

Global issues such as climate change, deforestation, loss of biological diversity, and destruction of the ozone layer potentially involve actors at every level, each with its own agenda. Unlike bilateral or regional issues, global programs require the participation of a large number of countries, many of which may consider the issue a low priority and some of which may be tempted to take a "free ride," receiving the benefits of the cooperative action without making major changes in their own behavior. The lack of enforcement mechanisms at the international level makes it difficult to prevent or punish such free riders. As was demonstrated by the June 1992 Rio de Janeiro UN Conference on Environment and Development (UNCED) and the follow up July 1997 New York conference ("Rio plus 5"), contention between rich and poor countries over what the major problems are, who caused them, and who should pay will continue to dominate multilateral efforts for some time to come.

The Roads to and from Rio

The participation of more than 160 nations in the UNCED indicated a general acceptance of the concept that even local and regional environmental problems often have global implications and that a multilateral governance approach is often necessary to address various types of ecological degradation. UNCED also explicitly acknowledged the inseparability of environmental and development issues, and—by addressing the pervasive economic implications of ecological problems—introduced a more intense and high-profile level of politicization into environmental negotiations.

Five major multilateral environmental agreements were reached at UNCED; these were added to the dozens already in various stages of ratification and implementation.[3] The UNCED agreements are the products of some of the most contentious international negotiations to date in the environmental arena. The negotiation process was characterized by several overall trends: a general North-South debate framework; increasing regionalism; and the growing importance of NGOs.

The Rio plus 5 conference focused largely on lack of progress in implementing the UNCED agreements. The United States Congress, for example, is still reticent to adopt targets for greenhouse gas reductions in the face of an ever-larger body of sound scientific evidence that such targets are long overdue. Developing countries also complained in New York that despite commitments to financial assistance at UNCED, the industrialized world has failed to deliver the financial aid needed to implement UNCED agreements. This lack of concrete progress points out the continuing centrality of domestic political dynamics, as well as financial issues at all levels, in the international environmental arena.

North-South Debate

UNCED's stated aim was to find ways to integrate the dual issues of environment and development. The UNCED Secretariat had hoped that negotiations would produce blueprints for countries, rich and poor, to pursue economic growth without squandering natural resources and to reverse environmental damage without derailing development programs.

Although some progress was made, UNCED in general failed to effectively integrate environmental and developmental concerns. Many observers felt that both environment and development issues were perceived and approached very differently by the "Northern" industrialized countries and the "Southern" developing countries. The North tends to consider such global environmental problems as climate change and deforestation to be top global priorities, whereas the South emphasizes the more traditional "development" issues of poverty, access to clean drinking water and sanitation facilities, and programs for financial and technological assistance. These differing approaches made for a painfully slow and disorganized UNCED negotiation process, which sometimes erupted into heated accusations and mutual distrust.

Some NGOs and Third World countries argue that the North's own historical industrialization process was unchecked by environmental regulation and that the North is thus responsible for most global environmental problems. They charge that the North is now attempting to redefine these problems in order to share responsibility and cost with the agricultural South. Developing countries argue that if the industrialized world wants them to change their behavior in a way that will benefit the North (perhaps even more than the South), then the North must make available new and additional assistance of all kinds to facilitate the changes.

The counterposition of the North is that any agreement to limit greenhouse gas or ozone-depleting emissions is meaningless without participation of such large developing countries as India, China, and Brazil, who in the next ten years could surpass industrialized country emissions severalfold if unchecked. The North further argues that the South also will benefit from such agreements and from efforts to pursue future economic development plans in an environmentally sound manner. While willing to offer some financial and technological assistance to developing countries for implementation of global environmental agreements, the industrialized countries feel that the expenses they will incur in changing their own behavior will be so great that they are simply unable to offer unlimited amounts of aid.

In its most antagonistic form, the North-South debate has been characterized by mutual suspicions: a "Southern" suspicion that the North is attempting to limit developing countries' sovereignty over their own territories and resources and to sabotage their development plans; and a "Northern" suspicion that the South is only engaging in environmental discussions in order to obtain huge additional amounts of financial aid and access to free technology. Although some on both sides do harbor these suspicions, most participants and observers would find them flawed for several reasons: Many people in developing countries are concerned with global environmental problems; industrialized countries also consider economic development a top priority, not only for the South but also for themselves; and the "North" and the "South" do not actually exist as cohesive, exclusive blocs with common positions.

Regionalism

The categories of "North" and "South" should be understood as loose groupings of countries with widely varying orientations. Some of the most heated debates in the UNCED process took place among the industrialized OECD countries. Disagreement between the United States and several European countries over the inclusion of targets and timetables for greenhouse gas reductions in the climate change convention almost derailed pre-UNCED negotiations. Disagreement among developing countries is equally pronounced; Brazil, for example, now favors global forest conservation efforts whereas others, such as Malaysia, have denounced such programs, viewing them as northern schemes to take control of southern resources.

The fall of the totalitarian governments of the former Soviet Union and several Eastern European countries has also created a new geographic/economic category that is neither North nor South. In UNCED documents, these countries are referred to as "economies in transition" that fall somewhere between the conditions and expectations placed on developed and developing countries.

It may be more accurate to view UNCED negotiations as among several single-issue regional coalitions. The groupings are small, the issues are fluid, and both are based on common geopolitical interests. For example, several South Pacific countries concerned about potential sea level rise formed the Association of Small Island States (AOSIS) to present a unified call for strict and binding limits on greenhouse gas emissions. Similarly, oil-producing Persian Gulf states also acted as a regional bloc in the climate negotiations, lobbying against any calls to limit carbon dioxide emissions. Because small numbers and proximity are more likely to engender real common interests than are large and disparate groupings, future negotiators may find regional cooperation efforts more efficient and effective than global agreements in the UNCED tradition.

NGOs

Most environmental NGOs are focused at the local and national levels, but international affiliations of these organizations are also becoming more widespread and active as concern for global issues grows. Increasingly, NGOs are organizing in developing countries, participating in international organizations, and proposing and negotiating international agreements. International level issues are becoming the top priority for some NGOs; Greenpeace recently announced it would be closing its domestic U.S. offices due to financial difficulties and will be conducting its business solely through its international network. More than one thousand NGOs participated in the "Global Forum," a parallel nongovernmental conference held in tandem with UNCED. These groups included human rights and indigenous peoples organizations in addition to specifically environmental NGOs; their constituencies ranged from single neighborhoods to worldwide membership.

NGOs played an important role throughout the almost five-year negotiating process leading up to the adoption of the climate change convention at UNCED. In 1990, for example, Greenpeace published a scientific report that paralleled and severely criticized a document issued by the UN-sponsored Intergovernmental Panel on Climate Change. The Greenpeace report generated much media and public attention on the climate change issue and the activities of the UN and national negotiating parties. Although this attention created some pressure upon governments, the final climate change convention ultimately fell far short of Greenpeace's findings and prescriptions.

At UNCED, NGOs discovered new ways to exert influence. They lobbied sympathetic national governments to sponsor positions and to present them as their own in official negotiations. Several of the amendments offered by the Australian, New Zealand, and some European delegations during the UNCED proceedings originated with NGO proposals. In addition, many NGOs prepared comments and alternate text to official UNCED documents under discussion and made them widely available; some of these NGO texts found their way into official discussions and often into the final products, despite the NGOs' official "observer" status. Numerous press conferences held at or near the UNCED and Global Forum sites increased the NGO-generated pressure on participating governments.

In some cases, the expertise and institutional memories of NGOs are resulting in their performing an indispensable staffing function for international environmental agreements and organizations. For example, the international NGO TRAFFIC (Trade Records Analysis of Flora and Fauna in Commerce) has come to play an integral role in the functioning of the CITES (Convention on International Trade in Endangered Species) treaty organization, with TRAFFIC studies and analyses contributing significantly to both the organizational decisionmaking processes and the media coverage of them.

Environment and Development

UNCED's title shows its explicit recognition of the dual nature of environmental and developmental issues. But the relationship between the two functions is multifaceted and complex. Environmental degradation can result both from affluence and from poverty, and countries approach issues of environment and development from the vantage points of their own levels of industrialization and environmental degradation.

Although poor countries often charge that the industrialized North is the source of most global atmospheric pollution, they cannot deny that much of the world's deforestation and consequent soil erosion, desertification, and flooding—resulting from widespread demand for fuelwood and poor agricultural practices—occurs within their own countries. Because industrial plants in Third World countries are often technologically outdated, they often lack environmental protection equipment readily available to their richer counterparts.

A perceived trade-off between forwarding environmental and developmental agendas is not only a phenomenon of the developing world. Just as developing countries argue that they can ill afford costly environmental protection laws and programs when their first priority is to improve the lives of their people, industrialized countries often cite similar arguments. Current economic problems make even rich countries feel they are too poor to implement costly environmental programs at home, let alone to fund other countries' participation in multilateral agreements.

In 1987, the UN-sponsored World Commission on Environment and Development (WCED) gave wide exposure to the concept of "sustainable development" in order to debunk the myth that a trade-off between environmental and developmental objectives must exist.[4] It suggested that environmentally sound development can actually enhance economic growth efforts and that, conversely, industrialization without environmental safeguards will be costly in the long run.

Sustainability

Sustainable development is a vaguely defined concept, requiring policymakers to take a long-range view that may entail some short-term sacrifice. Nonetheless, in the five years since UNCED, the concept has caught on widely, permeating the thought and planning processes of not just government agencies but corporate boardrooms, academic institutions, and citizens' groups alike. There is basic support for the idea that economic activity, environmental protection, and even social justice can be mutually reinforcing rather than competing endeavors. This represents great progress over the decade since the widespread introduction of the concept with the 1987 publication of *Our Common Future* by the WCED.

But it may be another five or more years before the acceptance of the sustainability principle is translated into concrete policies and practices, particularly at the national level. This comprehensive approach entails high costs and the alienation of some powerful interests in order to effect real changes. Just as with the UNCED process, these will continue to be the stumbling blocks of the future.

In addition, the sustainability concept, precisely because it is vague, comprehensive, and dynamic, is vulnerable to cooption by forces that would benefit by a lifting or softening of existing environmental laws and regulations (as well as to the *perception* by those favoring the preservation of existing regulatory approaches that such cooption is occurring). An example is the debate in CITES over the appropriate approach to regulation of the ivory trade and its relationship to the status of elephant populations.

Although the traditional approach of CITES to protecting wildlife species has been to ban or restrict commercial trade in endangered and threatened species and their derivative products, sustainable use advocates assert that in many cases wildlife can best be conserved by exploiting it for economic gain.[5] CITES imposed a trade ban on ivory in 1992 after the release of data indicating an alarming decimation of the African Elephant population over the 1980s. Since that time, the sustainable use approach has been forwarded in the context of the CITES elephant debate by pro-trade states, who view the ivory trade ban as evidence of an outdated, ineffective preservationist approach, dominated by scientists, government officials, and nongovernmental environmentalists in rich, industrialized countries. The "preservationist" approach is seen by many in the developing world as a means to a new imperialism, by which these elites can use institutions such as CITES to dictate to others what they can and cannot do with their own resources.

At the same time, some CITES and industrialized country government players view the sustainable use advocates and their argument with suspicion, believing that their own position has been unfairly caricatured into a "straw man" for the purpose of the advancement of an economically driven pro-trade position, regardless of the effect of such policies on wildlife conservation. In 1997, CITES accepted the sustainability approach and conditionally lifted the trade ban for three southern African countries on a trial basis.

This is a fine line, indeed, and only time and precise monitoring and evaluation of new policies will tell whether "sustainable" approaches to economic and environmental activity have a net positive effect on both arenas. For this reason, it is important to put in place precise systems of indicators and evaluation processes, with good baseline data, alongside new policies and practices.

"Politicization"

The international environmental negotiation process has been criticized in recent years for becoming increasingly politicized, as opposed to being a more objective scientific data-driven model of decisionmaking. The charge of "politicization" implies two assumptions, both false. It first implies that the process of international environmental negotiation has previously taken place in a vacuum, unaffected by the political concerns that drive actors in other issue areas. This assumption derives from the naive belief that governments are able to rely solely on "objective" scientific testimony to arrive at universally acceptable prescriptions. But environmental science is a complex body of disciplines in their relative infancies, and the scientific community itself is in a state of intense debate over ecological systems and processes. There is no indication that definitive scientific answers will be found in the short run, and this makes the science/policy relationship an extremely complicated one. Although close consideration of current scientific findings is a necessary component of sound environmental policy making, assumptions that either the findings themselves or the outcomes of the policy process will be shielded from the political milieu in which they exist is wishful thinking.

The second false assumption implied by the "politicization" criticism is that it is impossible to forge effective agreements within the political context of international relations. The international negotiating process, on any issue, will always be influenced by the political agendas of the participating countries. The domestic/foreign policy interface is a porous membrane in all types of governmental systems.[6] Policymakers in democracies and centralized systems alike are influenced by demands from domestic constituencies for competing foreign policy options. Negotiations would be better served if participants together sought goals and implementation strategies that satisfied both domestic and international agendas. A global emissions-trading program, for example, could give financial incentives to private actors to reduce pollution.

International Environmental Agreements in Context

A multilateral agreement, once reached, is only the midpoint in a process that begins well before the negotiation process starts and continues through steps toward implementation, creation of institutional frameworks, monitoring, enforcement, and modification. The latter steps are complicated by several general characteristics of international environmental cooperation efforts: participation is largely voluntary; implementation is costly for participating governments; there are often no real mechanisms capable of forcing compliance by signatories; and implementation and enforcement take place in the context of underlying political themes, agendas, and controversies.

These characteristics pose strong obstacles to the universal implementation of environmental agreements, but they do not doom such agreements to being only symbolic gestures. Although the creation of supranational institutions with real enforcement powers is unlikely in the foreseeable future, there are nonetheless powerful incentives for governments to forge and comply with multilateral environmental cooperation efforts. These incentives include the linkage of environmental cooperation to other types of international relations, including economic assistance, trade, technological cooperation, and security relationships. The strong presence of NGOs and the media can focus public attention on "environmental outlaws." The high profile of environmental issues on most domestic political agendas encourages national leaders to take the "high road" on the international environmental stage. More emphasis will likely be placed on these "carrots" in the negotiation and implementation of future agreements.

Outlook

Perhaps more than any other single issue in international relations, environmental problems point out the limitations of the nation-state as an international actor. Disputes over natural resource exploitation, transboundary pollution, and energy generation all have at their base a classic collective action dilemma. Even where there is a universally acknowledged common interest among members of a group, "unless there is coercion or some other special device to make individual [countries] act in their common interest, rational, self-interested individuals will not [usually] act to achieve their common or group interests."[7] The challenges for international environmental policy makers involve identifying the common interest, devising a widely accepted strategy for advancing that interest, and providing the "special device" that will motivate countries to voluntarily implement that strategy in concert, but ultimately, each within its own unique constraints.

Although global conventions and other multilateral cooperation efforts are essential for providing a "level playing field" on which to take otherwise unilaterally costly action, the international community may find that post-UNCED achievements can be more easily accomplished at the local, national, and regional levels. Smaller-scale projects are less susceptible to such UNCED-type pitfalls as emphasizing symbolism over substance, succumbing to global ideological debates, relying on the unlikely transnational transfer of large-scale financial or technological assistance as a contingency for implementing agreed-upon principles, and sacrificing commitment to real action for achieving consensus on text.

Unilateral actions, such as the introduction by some European countries of a carbon tax—well ahead of an anticipated but stalled EU-wide tax—have the benefit of being less complicated and, often, less costly than the implementation of a new global convention. Regional and bilateral environmental assistance programs are implemented at a lower level on a more routine basis than global agreements and may benefit from a relative lack of public and media scrutiny. Programs involving the participation of only a few actors can be put into practice relatively quickly, rather than being dragged through the signing, ratification, and other entry-into-force requirements that can slow global conventions. Examples include Norway assisting Russia to clean up radioactive waste sites and the United States helping Mexico to implement a border-area environmental cleanup plan.

Such projects are generally designed to advance the national interests of the country or countries involved, so the collective action problem of national interests conflicting with common interests often does not come into play. It is the local and regional environmental problems that, in aggregate, make up most global problems. Relatively small improvements made at the local level, therefore, also contribute in aggregate to global solutions. As long as sovereign nation-states are the principal actors in the international system, the forging of global agreements will not be a panacea or a substitute for ongoing unilateral environmental activities but rather will complement them.

Notes

1. See William B. Wood, George J. Demko, and Phyllis Mofson, "Ecopolitics in the Global Greenhouse," *Environment* 31 (September 1989): 12–17, 32–34, for a more in-depth discussion of the ecopolitical hierarchy.

2. Murray Feshbach and Alfred Friendly, Jr., *Ecocide in the USSR: Health and Nature Under Siege* (New York: Basic Books, 1992), pp. 2–3.

3. The UNCED agreements are the Rio Declaration on Environment and Development, the United Nations Framework Convention on Climate Change, the United National Framework Convention on Biodiversity, a nonlegally binding authoritative statement of principles for a global consensus on the management, conservation, and sustainable use of all types of forests, and Agenda 21.

4. World Commission on Environment and Development, *Our Common Future* (Oxford: Oxford University Press, 1987), p. 43. The report defines sustainable development as "development that meets the needs of the present without compromising the ability of future generations to meet their own needs."

5. Phyllis Mofson, *The Behavior of States in an International Wildlife Conservation Regime: Japan, Zimbabwe, and CITES*. Ph.D. diss., University of Maryland at College Park, 1996, pp. 73–76.

6. Robert D. Putnam, "Diplomacy and Domestic Politics: The Logic of Two-Level Games," *International Organization* 42 (Summer 1988): 427–460.

7. Mancur Olson, *The Logic of Collective Action* (Cambridge: Harvard University Press, 1965), p. 2.

References

Boardman, Robert. *International Organization and the Conservation of Nature.* Bloomington: Indiana University Press, 1981.

Cairncross, Frances. *Costing the Earth*. Boston: Harvard Business School Press, 1992.

Carroll, John E., ed. *International Environmental Diplomacy*. Cambridge: Cambridge University Press, 1988.

Convention on Biological Diversity. United Nations Document UNEP/Bio.Div/ Conf/L.2, 1992.

Feshbach, Murray, and Alfred Friendly, Jr. *Ecocide in the USSR: Health and Nature Under Siege*. New York: Basic Books, 1992.

Framework Convention on Climate Change. United Nations Document A/AC237/18, 1992.

Haas, Peter M., Marc A. Levy, and Edward A. Parson. "How Should We Judge UNCED's Success?" *Environment* 34, no. 8 (October 1992).

Non-legally Binding Authoritative Statement of Principles for a Global Consensus on the Management, Conservation, and Sustainable Development of All Types of Forests. United Nations Document A/CONF.151/6/Rev.1, 1992.

Olson, Mancur. *The Logic of Collective Action*. Cambridge: Harvard University Press, 1965.

Parson, Edward A., Peter M. Haas, and Marc A. Levy. "A Summary of the Major Documents Signed at the Earth Summit and the Global Forum." *Environment* 34, no. 8 (October 1992).

Pirages, Dennis. *Global Technopolitics: The International Politics of Technology and Resources*. Pacific Grove, CA: Brooks/Cole, 1989.

Rio Declaration on Environment and Development. UN Document A/CONF.151/ PC/WG.III/L.33/Rev. 1, 1992.

UNCED *Agenda 21*. United Nations Document, 1992.

Wood, William B., George J. Demko, and Phyllis Mofson. "Ecopolitics in the Global Greenhouse." *Environment* 31, no. 7 (September 1989), 12–17, 32–34.

World Commission on Environment and Development, *Our Common Future*. Oxford: Oxford University Press, 1987.

Young, Oran. *International Cooperation: Building Regimes for Natural Resources and the Environment*. Ithaca: Cornell University Press, 1989.

Chapter Fourteen

Nongovernmental Organizations on the Geopolitical Front Line

MARIE D. PRICE

Much is written about the relative decline of the state in an increasingly globalized world. At one level multinational corporations, financial institutions, and supranational entities such as the United Nations seem to be usurping the de facto sovereignty of states. At the same time, subnational groups divided along ethnic, sectarian, or ideological lines may also be seeking to undermine the functioning integrity of individual states, as witnessed in Sri Lanka, Rwanda, Myanmar (Burma), and the former Yugoslavia. The media tends to dwell on the *centrifugal* forces that challenge a state's centralized authority and neglect the considerable *centripetal* forces that hold states and their citizens together. Even in the context of the post–Cold War era, states still have tremendous power. As far as international affairs and geopolitics are concerned, states and national governments continue to be a primary locus of study. But the era of the state's dominance over political geographic analysis is over.

A relatively new category of geopolitical player merits attention; nongovernmental organizations (NGOs) have been increasingly effective in negotiating for political space within a state-based world system.[1] Their growing numbers and popularity suggest that they are poised to be still more influential in the next century. These so-called nongovernments are not anti-government, in fact, many maintain close relationships with the national governments with whom they operate. Yet they are markedly different from state governments in their responsiveness, flexibility, and ability to operate at various scales. Some NGOs are major international organizations with multimillion dollar budgets (International Red Cross, CARE), and others are community-based self-help groups relying on a few volunteers. Community-based organizations may form to clean up a park, provide credit to micro-enterprises, build an irrigation canal, or set up a soup kitchen. But even these locally oriented groups may receive technical or financial support from larger international NGOs. Thus, the reach of these organizations is not confined to state boundaries but can be binational, international, or even global. Also, with breakthroughs in telecommunications, NGOs have become effective at diffusing information and publicizing their causes. The NGO-driven effort against the use of land mines, for example, was largely waged through faxes and the Internet with an organizational base in Putney, Vermont.[2]

NGOs are creating a significant place for themselves in the constellation of political forces reordering our world. This chapter begins with a working definition of NGOs and then examines the interaction between NGOs, states, and intergovernmental agencies such as the United Nations. Many of the oldest NGOs had their start as charitable organizations concerned with humanitarian work in war-torn or famine-weary lands. Yet the 1980s marked the decade when NGOs emerged out of the shadows of grassroots action and into the spotlight of world political debate. Over the past two decades, thousands of NGOs have formed, representing an important institutional shift in how people mobilize to better their lives and effect change. Although NGOs have played a vital role in the economic development of certain localities, their geopolitical impact has been greatest on humanitarian relief and environmental conservation. Examples from these two fields are highlighted to show what NGOs are doing and how they are influencing world politics. This chapter concludes with a discussion of NGO distribution and the benefits and drawbacks of nongovernmental movements.

What Is an NGO?

The term NGO is a catch-all phrase that is as popular as it is vague.[3] Generally speaking, these are private, not-for-profit entities that provide services, particularly to marginalized or underrepresented groups. NGOs are involved in humanitarian relief, primary health care, community development, advocacy, education, and environmental conservation. Although there are numerous subcategories of NGOs, there is an important division between grassroots organizations and larger supporting institutions. Grassroots organizations (GROs) or community-based organizations are created and sustained by the people they serve. Small by their nature, they often form linkages with other GROs and with larger NGOs that can help them. Grassroots support organizations (GRSOs) or international NGOs are often based in the developed countries and provide technical, organizational, and financial support to local groups. Save the Children Federation, for example, has offices in dozens of countries. From these centers staff and resources are dispatched to communities in need, usually through grassroots organizations. GRSOs tend to be staffed with professionals from the developed world and financed by private donors or foundations; the communities they serve are usually not members of the organization.

The key operational feature that distinguishes an NGO from a governmental organization is that it is *privately run*. Managed by a board of directors or trustees, NGOs are accountable to their donors and to the particular communities they serve. GROs are often run by the members themselves, who in turn rely upon the organization's services. Here accountability is to its members and not necessarily to the community at large. On closer inspection, however, the private/public line is frequently blurred. For example, some of the largest NGOs in the United States (CARE, the Red Cross, and Catholic Relief Services) receive support from the U.S. government through the Food for Peace program established by Public Law 480. Millions of dollars in excess foodstuffs, mostly grains, are given to these organizations for distribution in developing countries. The U.S. government also either transports these materials or pays the transportation costs. This special relationship provides crucial financial support to the work of these organizations, along with furthering the international objectives of the U.S. government. Yet the NGOs involved insist that their autonomy is not compromised by this private-public partnership.

With the rising popularity of NGOs in the developing world, the division between public and private organizations becomes even more murky. Impressed with the ability of NGOs to solicit international funds and technical support, many governments have created NGO affiliates for their various ministries. These quasi-NGOs exist as a conduit for international grants and donations as well as a means to deliver semi-privatized services. Thousands of GROs in Egypt, for example, receive state financial support and are partially controlled through a government ministry. Although the majority of NGOs in the developing world are technically independent from the state, it is not unusual for these groups to register with state agencies and even be taxed. In addition, extremely successful NGO projects may later become the prototype for state-directed initiatives.

The idea that "private" NGOs are *nonprofit* separates them from *for-profit* businesses. This distinction is important because the nonprofit status implies that an organization is engaged in an altruistic provision of a service for little or no cost. This gives NGOs a mantle of neutrality that is especially important when they work internationally and in conflict zones. The disadvantage of a nonprofit status is that NGOs must rely on donations or government contracts, and they often struggle to be financially self-sustaining. GRSOs are constantly appealing to foundations, private donors, and in some cases governments to support their goals. Typically, money is drawn from the developed world and channeled to the developing world. Yet southern NGOs are wary of dependence on northern organizations. Some have sought domestic sources of funding from members, businesses, and government agencies. Others charge for services to cover expenses but not enough to make a profit. Ironically, the search for economic stability sometimes forces NGOs to be more accountable to their donors than the constituents they serve.

NGOs claim to serve the needs of marginalized or underrepresented peoples; this is especially true in the developing world. Their ability to serve and even empower the less powerful is perhaps the major reason why nongovernmental approaches to coping with poverty have become so popular. If state governments tend to respond to the demands of the privileged or the majority, NGOs typically serve ethnic or religious minorities, the poor, and tens of millions of internally and externally displaced peoples fleeing environmental degradation and/or political turmoil. When not directly service oriented, many NGOs are advocates for reform, publicizing the ineffectiveness of state policies with the hope of making governments more accountable. Consequently, NGOs and states are typically entangled in a reluctant partnership, recognizing the advantages of cooperation but not totally trusting each other's methods and motives. The goal of NGOs is not to replace government but to work with government so that basic needs are met and policy reforms are made. Those who support the work of NGOs in the developing world also believe that NGOs are excellent mediators between the individual and the state and thus promote political and social development. By building the foundations of civil society, NGOs make states more accountable and responsive to their citizens. This complex relationship between NGOs and states deserves closer inspection.

The Rise of Nongovernments in a State-Based System

Given the many financial and political obstacles that NGOs face, why have they become so numerous? Today there are more than 200,000 GROs working to develop their own communities in Asia, Africa, and Latin America. On top of that, some 50,000 supporting NGOs (GRSOs) are active in the developing world.[4]

The popularity of nongovernmental solutions to social and environmental problems results from a convergence of ideologies and organizational practices. In the 1960s, both religious and secular organizations promoted the ideas of cooperatives and community self-help groups. This organizational base was elaborated on over time as citizens became increasingly frustrated with the inefficiency and unresponsiveness of government institutions in dealing with chronic problems such as poverty, inadequate education, and undrinkable water.

Further support for NGOs came from neoliberal ideologies popularized in the 1980s. Neoliberalism espouses the superiority of the private sector over the public sector to address certain social and economic problems. Accordingly, NGOs are called upon to provide services because they are cheaper, more efficient, and better integrated with the constituents they serve. Today, the familiar organizational structure of NGOs coupled with a preference toward private approaches have made NGOs a common part of the landscape in the developing world. As these organizations have evolved, alliances have formed with other NGOs as well as with their host governments, multilateral lending agencies, and intergovernmental organizations. In the North, NGOs are viewed as more efficient and less corrupt than bloated government bureaucracies. The view from the South is that NGOs are more flexible and responsive to basic human needs.

This should not imply that NGOs are universally welcomed or tolerated by states. NGO-state relations are not easily generalizable. Much depends on the political context within a particular state; the more authoritarian the government, the less likelihood of a flourishing NGO sector. A few regional patterns do exist. Throughout Latin America NGOs have been active since the 1960s and have a tradition of being independent from the state. Even when Brazil, Chile, and Argentina had authoritarian military governments in the 1960s and 1970s, NGOs were tolerated—provided they maintained political neutrality. Local NGOs in Asia, by contrast, evolved under stricter government regulation. Government officials typically insist that NGOs register and participate in collaborative enterprises with state agencies. Although this can foster beneficial coordination, it can also limit an NGO's autonomy. Asian NGOs are publicly appreciated for the valuable services they provide in the fields of public health, family planning, and rural development. Yet those that openly clash with government policies, such as environmental and human rights NGOs, are quickly suppressed. India leads Asia in the number of NGOs with more than 12,000; there state and local governments are generally supportive of the services NGOs provide, and NGO-state collaboration is common. By contrast in China, privately run domestic NGOs are virtually unheard of, although the government began allowing in foreign NGOs about a decade ago.

In Africa, indigenous NGOs tended to be small and low profile until the 1980s. Today, most African governments are aware of these organizations but exert mixed policies toward them. South Africa has the largest network of NGOs, which were nurtured during the country's long struggle to end apartheid. In Kenya, NGOs are tolerated, yet groups that aggressively advocate political reform or human rights have to go underground. In Africa, as elsewhere, attempts to control or regulate NGOs are often guided by a fear of foreign influence. The 400 NGOs in Kenya are collectively the largest source of foreign exchange in the country, which generates suspicion and jealousy on the part of state officials.[5]

State-NGO relations can be neutral, collaborative, or coercive. Arguably, coordination of public and private services can lead to more effective coverage. Forced to adopt structural adjustment policies, many governments have found GROs and GRSOs to be important instruments for helping those most negatively affected by these changes, the poor. The Bolivian government, for example, has channeled millions of dollars directly to GROs involved in health and education sectors through its Bolivia Social Investment Fund. This fund, partially financed through a World Bank loan, uses grants to finance hundreds of small-scale projects each year.

In general, a pattern of more interorganizational collaboration, rather than less, is becoming common. But that does not dilute the resentment felt by some state agencies toward NGOs. The chief complaints are that NGOs are too heavily influenced by international NGOs, foreign capital, and "Western" concerns. There are differences in resources as well. It is not unusual for an environmental NGO to have better access to computers, GIS, and satellite images than many government ministries. In extreme cases, NGOs are viewed as neocolonial agents pushing the agenda of the North on unwitting GROs in the South; such fears persist because NGOs function as part of an international network.

The Global Reach of NGOs

Another way to understand the increased geopolitical influence of NGOs is through their evolution. David Korten uses a generational model to distinguish today's third-generation NGO from first- and second-generation ones. First-generation NGOs emerged more than a century ago and were concerned with relief and welfare. The oldest humanitarian relief group is the Red Cross, founded in 1863 by a Swiss businessman. Known today as the International Committee of the Red Cross, this NGO has been formally assigned to assist victims of civil conflict by the Geneva Convention. It also is the only humanitarian group with a mandate to monitor the treatment of prisoners of war. To assist in relief operations the ICRC typically relies on 160 separate national Red Cross programs as well as a coordinating body called the International Federation of Red Cross and Red Crescent Societies (IFRC). The American Red Cross, founded by Clara Barton in 1881, is part of this federation. Save the Children was started in the 1930s and Catholic Relief Services began in the 1940s in response to the pressing humanitarian needs caused by World War II. Geared to providing relief for political refugees and victims of natural disasters, they continue to serve millions in need. Although these groups are less inclined to promote long-term sustainable solutions, many have added a development component to their general mission of relief.

Second-generation NGOs began to emerge in the 1960s in response to the gaps left by first-generation groups. These groups typically focused on development at the local level. Such community-based self-help groups sought to improve local infrastructure, access to capital, and health care. Small in size and local in impact, thousands of these groups formed in the developing world to better community living conditions. They rarely attempted to address the structural forces of poverty and inequality but aimed to improve basic needs. The organizational base for many current GROs came from this second-wave NGO philosophy that gives priority to self-sufficiency and local action.

What distinguishes third-generation groups is a specific intent to reshape public policies and attitudes that perpetuate poverty, social injustice, and ecological devastation. This goal requires multilateral engagement and gives third-generation groups a long-term and broad-based purpose that goes beyond immediate relief or community development. The NGOs themselves do not have to be large to do this, only part of a network that provides a forum to express their needs, accomplishments, and vision. Even traditional NGOs have been influencing public policy in particular countries for decades—a good example of this is the widespread adoption of family planning practices promoted by offshoots of the International Planned Parenthood Federation—but since the 1980s NGOs have received greater recognition from intergovernmental agencies, and their access to the media and telecommunications has greatly improved.

Nowhere is this outreach more obvious than the collaboration between the United Nations and NGOs. The members of the United Nations are, of course, states. Yet NGOs continue to work closely with relevant UN bodies in the areas of relief, development, and environment. The full extent of this collaboration was evident at the UN Conference on Environment and Development in Rio de Janeiro in 1992. What was popularly termed the Earth Summit attracted some 9,000 environmental NGOs, which brought 22,000 representatives. The Population Conference in Cairo (1994), the Women's Conference in Beijing (1995), and the Climate Change Conference in Kyoto (1997) had equally high levels of NGO participation. Although they meet separately from the actual meeting of member states, these joint conferences allow NGOs to voice their concerns in a meaningful way to a world audience. Beyond conference attendance, the UN actually works with hundreds of NGOs in its humanitarian relief missions—a partnership that is discussed later.

Another breakthrough for third-generation groups was the decision taken by the World Bank in the 1980s to work directly with NGOs in operational and policy areas, especially in the fields of social issues and the environment. The World Bank felt that in some cases NGOs could provide more rapid assessment and better local contacts than governmental ministries. In 1982 the NGO–World Bank Committee was established "to strengthen relations and expand operational cooperation between the Bank and NGOs."[6] Officially the World Bank makes loans to states, but increasingly services are contracted directly to state-approved NGOs that can better implement or assess a project. Interestingly, the bank's decision to be more inclusive was due, in part, to the pressure NGOs placed on the bank for its failure to address the environmental consequences of its loans. The NGO-led fight against road and dam construction in tropical forests, for example, led the World Bank to include environmental impact assessments in their decisionmaking process. Smaller lenders, such as the Inter-American Development Bank, have also begun to include environmental variables when considering loans.

Case studies of NGO activities in environmental efforts and humanitarian intervention illustrate the increased effectiveness of NGO-led initiatives. Although NGOs are not the only players in these fields, their impact has been magnified by their willingness to coordinate with others. And they are probably the most effective organizational tool for politically marginalized groups whose voice would otherwise go unheard.

NGO Environmental Leadership

Environmental concerns have fostered many forms of activism, both public and private. Yet environmentally oriented NGOs are one of the most important participants in an emerging "ecopolitical hierarchy" that is pushing for ecological accountability. At the base of this hierarchy are GROs and GRSOs working at local, regional, and national levels to insure that environmental issues within particular states are addressed. At the upper levels, international environmental NGOs, such as Green Peace, World Wildlife Federation, and Friends of the Earth, participate in forums to address global environmental issues. Working alongside intergovernmental and governmental bodies, these major NGOs have participated in international negotiations for greenhouse gas reduction, trade bans on endangered species, and conservation of biological diversity.

Emerging international environmental coalitions that monitor governments and industries and encourage ecologically sound policies are playing key roles at different levels of the ecopolitical hierarchy. The Japan Tropical Forest Action Network was founded by ten Japanese NGOs in the 1980s but now has a network of participating organizations in Asia, North America, Europe, and Latin America. Another organization, Tiers-Monde (Environment and Development in the Third World), formed in West Africa but now has offices throughout Africa as well as the Caribbean and India. This particular consortium began with UN seed money and is now financed by European governments. The European Environmental Bureau is a coalition of more than one hundred environmental groups from twenty European countries. Tracking environmental provisions set by the European Union, this group has access to the European Commission and represents European environmental NGOs at international meetings.[7]

The strategic advantages of such coalitions are obvious. By forming umbrella organizations, like-minded groups can express strongly their interests at major international conferences. Likewise, such a network keeps groups in contact with each other and avoids needless duplication of effort. Moreover, since many environmental problems are transnational in scope, the formation of regional environmental coalitions fosters international dialogue and cooperation. At the national level, environmental groups have also formed coalitions to lobby their governments more effectively. The Mexican Conservationist Federation, an umbrella group that formed in 1985, pushed Mexico's political parties to take environmental positions on issues affecting the country's soils, water, air, and forests. Prior to FCM's efforts, environmental issues were virtually ignored in the nation's political discussions.

Environmental NGOs now work in virtually every country in the world, a presence not found even a decade ago. In this global diffusion, U.S. environmental groups have had the greatest influence on the promotion of environmentally sensitive development. The combination of their experience, access to funding, and support at home have helped them to build a vast international network. Some NGOs, such as the Sierra Club and Audubon Society have existed for nearly a century; others, such as the Nature Conservancy, Friends of the Earth, and the National Resources Defense Council, have two or three decades of policy-influencing experience. Although their original focus was domestic, the global dimensions of environmental problems have inspired their international outreach. By the 1980s the practice of supporting start-up or sister organizations in the developing world was firmly established.

Many fledgling environmental groups in the developing world owe their existence to start-up funds (sometimes as little as $5,000 to $15,000) and annual grants from international NGOs based in the United States. Some of the largest transfers of capital were from U.S. NGOs to Latin American NGOs. In the 1980s, for example, the World Wildlife Fund gave $25 million to environmental NGOs in the region. Similarly, in the 1980s the Nature Conservancy developed its Parks in Peril program, which channeled millions in grant dollars to environmental NGOs that worked closely with national parks in Latin America. Other philanthropic foundations, such as the MacArthur Foundation, the Pan-American Development Foundation, and the Ford Foundation, also funded environmental work in the region.

This transfer of "green funding" from North to South unleashed a debate among those following new social movements. Was an environmental agenda being imposed upon the developing world from the wealthier North? No one doubts that the world's poor usually consider short-term economic survival over long-term environmental sustainability. Yet it does not mean that they are unaware of how environmental degradation affects their daily lives. On the contrary, NGOs in countries such as Brazil and India sustained their own environmental agenda long before foreign institutional support was widespread. In some cases, the infusion of foreign support triggered the development of environmental groups. Yet these organizations are only sustainable when there is a local base of support for their work. Consequently, there are philosophical and practical differences between northern and southern environmental groups. Nature conservation is stressed in the North, whereas in the developing world NGOs must integrate the goals of environmental conservation, social justice, and sustainable development.

To demonstrate how this complex web of financial and institutional support works, the case of a Paraguayan NGO is instructive. Survival-Friends of Paraguay *(Sobrevivencia)* was founded in 1986 in Asunción by a few professional Paraguayans concerned with the health of the environment and the rights of indigenous and poor people. By 1992 the organization had an affiliation with U.S. based Friends of the Earth and a clear mission—defending those displaced by an internationally funded dam project on the Paraná River called Yacyretá, which began in 1979. Sobrevivencia did not create the dissatisfaction surrounding the resettlement aspects of the project; neighborhood groups had vigorously complained for years about being neglected after promises of resettlement. Sections of the town of Encarnación (population 30,000), for example, were inundated by the rising reservoir leaving neighborhoods cut off from each other. Village peasants were promised resettlement years ago and were told not to make improvements on their property. In the meantime, stagnant and polluted water from the reservoir contaminated groundwater supplies, incapacitated sanitation systems, destroyed crops, and killed fish. Complaints to local officials about this environmental degradation brought no improvement in living conditions. When the reservoir eventually fills to capacity, the island home of a Guarani indigenous tribe and parts of the city of Posadas will also be inundated.

As an internationally connected NGO, Sobrevivencia gave legal expression to peasant groups and directed their complaints into an international arena. Their strategy was bold in its simplicity. After familiarizing themselves with World Bank procedures, they filed a nineteen-page Request for Inspection report that resulted in the bank halting the rising waters of the dam until the claimants' concerns were addressed. Once the bank's inspection team was involved, the governments of Argentina and Paraguay had to comply with their findings. The difficulty now is to determine what the next step should be: Some want immediate resettlement whereas others want the dam's present water level to be maintained, thereby reducing the area of inundation but also limiting the dam's energy production. The most radical voices call for eliminating the dam entirely, an unlikely option since nearly $2 billion has been invested in the project. Although the process is now out of Sobrevivencia's hands, they are being besieged by requests from other environmental groups who would like to copy their legal strategy. Whatever happens to the village of Encarnación and its residents, Sobrevivencia's ability to spotlight the negative environmental and human costs associated with the dam has clearly demonstrated at the local, national, and international levels that a well-focused environmental NGO can influence economic decisionmakers.

The Humanitarian Outreach of NGOs

The history of NGO involvement in humanitarian crises spans more than a century. They are often the first groups to serve victims of war, famine, and natural disasters since they typically have a presence in the region long before UN or governmental relief services arrive. Catholic Relief Services (CRS), for example, supports development projects in dozens of Third World countries through its country-based affiliates called *Caritas*. When a crisis situation arises, CRS uses its local contacts, often a network of churches, to quickly distribute food, medicine, and other relief services. Other major NGOs, such as CARE, OXFAM, World Vision, and Save the Children, work directly with UN offices and governmental ministries to deliver vital goods and services, as witnessed in Rwanda, Somalia, Bosnia, and Zaire.

Since the end of the Cold War, NGOs have been called upon to do the work of states. This is especially true when dealing with growing numbers of refugees and internally displaced people. The U.S. Committee on Refugees estimates that there were some 33 million refugees and internally displaced persons in 1996. Most of these people are found in the developing world, moving within and between poor countries. Their need for outside assistance is tremendous. The single largest host of refugees is the country of Iran, providing a safe haven for more than two million refugees (mostly Kurds and Afghanis). Extremely poor African countries such as Guinea, and until recently Zaire, also host hundreds of thousands of refugees. NGOs are asked to step in precisely because their nongovernmental status lends them political neutrality. Moreover, the relief organizations themselves stress their humanitarian concerns and try to avoid taking political stands—something states rarely do.

Clothed in a mantle of neutrality, NGOs traditionally were spared direct attack by warring factions. Yet even this is changing, in part because humanitarian groups are being asked to serve in riskier and more complex humanitarian crises that involve the UN, governments, and the military. As the Somalian state collapsed into rival clans in the early 1990s, NGO staffs and their supplies were easy targets for the warlords who kidnapped relief workers and ransomed them for supplies. The International Red Cross experienced the worst atrocity ever directed toward its staff in 1996 when six hospital workers were gunned down in the middle of the night at a Chechen hospital. No faction claimed responsibility, yet the organization's belief in its universally accepted righteousness was deeply shaken. If humanitarian NGOs are to continue serving victims in conflict zones, they may have to work under the protection of UN peacekeepers or the military.

The humanitarian groups that most clearly influence policy and politics are those engaged in human rights. Their mission is expressly political, publicizing cases of state oppression of political dissidents and ethnic minorities. International organizations such as Amnesty International and Human Rights Watch have monitored state-sponsored oppression for decades, often providing the only outside interpretation of abuses. Rather than offering direct relief, their mission is to responsibly record violations of human rights in an effort to help punish those responsible and to bring about political and legal reform. For example, Human Rights Watch/Asia issued a 600-page report in 1994 documenting more than 1,000 political and religious prisoners in China and Tibet. Such NGOs, and not individual states, have been the main agents documenting human rights abuses and pressuring the UN and individual states to make governments abide by international standards of conduct. Although these NGOs have no enforcement capabilities of their own, they try to sway public opinion through information so that governments are pressured to reform themselves. Failing that, they can lobby the international community to impose sanctions on repeat violators of human rights. In that regard, the use of the Internet has been a valuable tool for NGOs to diffuse information quickly and inexpensively. Through the collaboration of human rights groups, indigenous groups, the UN, and signatory states there is now an International Declaration of Human Rights.

Indigenous peoples have also created NGOs to represent their collective political aspirations. They too have been active participants in UN conferences to voice the concerns of indigenous peoples, which are often ignored by national governments. Their arguments for better representation, semi-autonomy, and territorial control are often couched in the language of human rights, cultural survival, and environmental stewardship. Such groups can be national or international in scope. The Inuit Circumpolar Conference (ICC), for example, represents indigenous peoples in the Arctic region, including Inuits who reside in Canada, Russia, Greenland, and the United States. The ICC opposes militarization, encourages protection of native lands and cultures, and promotes self-government. Similar groups exist to represent the collective interests of native peoples in the Amazon Basin.

Sometimes indigenous NGOs are so successful in making their case for territorial and cultural control that they actually evolve into governmental bodies. The Inuit of northeastern Canada reached a settlement with the Canadian government that will award them a 136,000-square-mile homeland called Nunavut by 1999. The land comprises some 60 percent of Canada's Northwest Territories, and the 22,000 Inuit of the eastern Arctic will also receive mineral rights, a cash payment of nearly one billion dollars, and local autonomy while still being part of the Canadian state. This is the largest land settlement awarded to native peoples this century and is seen by many indigenous and human rights groups as a progressive and precedent-setting step.

The United Nations and NGOs

When the United Nations was chartered in 1945 its mission was to promote international peace and security and to help resolve humanitarian, economic, and social problems through member cooperation. For much of the organization's history there has been a reluctance to interfere with the internal affairs of member states. Yet the UN's role in the post–Cold War period is decidedly more interventionist. When states repeatedly violate the basic human rights of their citizens, especially minority groups, UN agencies have shown a greater willingness to assert themselves in internal conflicts. Most notable is the expanded role of the UN in leading complex humanitarian missions such as assisting the Kurds in northern Iraq and attempting to resolve ethnic fighting in Somalia and the former Yugoslavia.

UN membership is exclusive to sovereign states, but this intergovernmental body relies on partnerships with NGOs to carry out its mission. With the dramatic increase in complex humanitarian crises brought on by the end of the Cold War, this UN-NGO alliance is vital in extending the overall effectiveness and reach of humanitarian assistance. To coordinate UN and private efforts, the Department of Humanitarian Affairs was established in 1992. Several UN agencies, including the High Commissioner for Refugees (UNHCR), the World Food Program (WFP), the Children's Fund (UNICEF), and the World Health Organization (WHO) regularly provide humanitarian assistance. In turn, they rely on the talents and skills of several major NGOs, including the Red Cross, CARE, CRS, OXFAM, World Vision, Lutheran World Federation, Adventist Development Relief Agency, *Médecins Sans Frontières*, *Action Internationale Contre le Faim*, and Save the Children Federation. There are about 15 to 20 major secular or religious international organizations that regularly engage in relief work. In addition, scores of smaller national and local organizations participate in individual cases.

Initially the UN preferred working with internationally recognized northern NGOs to assist them with refugees, relief work, and development projects. By the 1980s, it became apparent that local NGOs offered important services that larger international organizations could not provide. In addition to having fewer cultural obstacles and better contacts, local NGOs often provide the same services for less money. Contracting directly with local NGOs has the added benefit of improving their institutional capabilities, which may serve to enhance civil society in their home country. Local NGOs engaged in development or human rights advocacy often receive their first sustained financial backing from the UN, allowing them to implement long-term projects. The UN's institutional and financial support of NGOs reflects the expanded role that private organizations now play in the developing world. In embracing the work of local NGOs, the UN has directly and indirectly nurtured the growth of this sector.

Mapping NGO Influence

In giving voice to the needs of marginalized peoples, NGOs have become a vital organizing tool for channeling resources and information and for effecting social and ecological change. Arguably, the presence of NGOs has increased levels of governmental responsiveness toward environmental problems. Although some of these responses may be of limited utility, a political dialogue has opened where once there was none. The long-standing humanitarian focus of NGOs has not wavered, even though their areas of interest have shifted in the post–Cold War world. They continue to serve those most in need: the poor, the stateless, and those displaced by war. As the institutional force of NGOs has grown, collaboration with government agencies, intergovernmental bodies, and other NGOs has become more common. Yet NGOs must still maintain their relative autonomy so that they can challenge state policies as warranted. Geopolitically, NGO networks are multiscalar, influencing local, national, and global decisions. Because of their multiscalar influence, their spatial distribution around the world invites further geographic inquiry.

The following maps offer a crude measure of the institutional pervasiveness of NGOs in the developing world. Figure 14.1 shows the number of NGOs per country.[8] Rather than a regional pattern, scattered countries show heavy reliance on NGOs. In Africa, South Africa leads the continent, but Egypt and Tunisia also have a vigorous NGO sector. Brazil, Colombia, and Peru have the most NGOs in Latin America, whereas the leading NGO countries in Asia are India, Bangladesh, Indonesia, and the Philippines. The

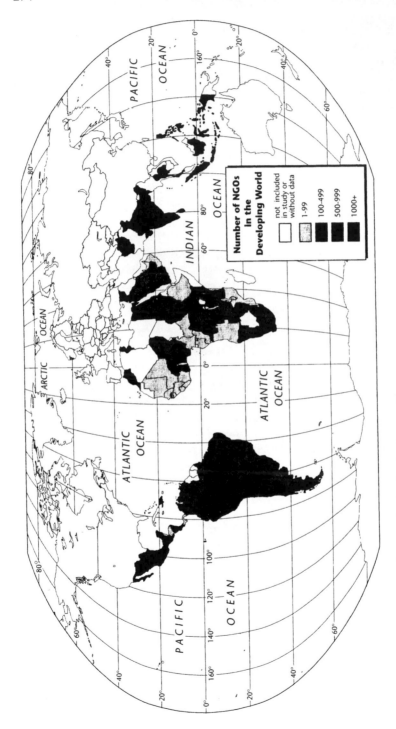

FIGURE 14.1 Number of NGOs in the developing world.

SOURCE: Julie Fisher, Nongovernments: NGOs and the Political Development of the Third World (West Hartford, CT: Kumarian Press, 1997), 162–164.

number of NGOs per one million inhabitants (see Figure 14.2), by comparison, offers a different pattern of NGO engagement. In demographically small countries, such as Lebanon or Palestine, there are more than 300 and 600 NGOs respectively per one million inhabitants. El Salvador, Namibia, Tunisia, and Zimbabwe have a greater NGO density than India with its 13 NGOs per million inhabitants.

Although these estimates are rough, NGOs appear to exist in nearly every developing country, and most have more than thirty such organizations. The obvious gap in the map is China where there is no government support for NGOs. Also, data for Central Asian countries are poor. Given their recent history as part of the Soviet Union, their experience with NGOs is limited but growing quickly. A map of international NGO support to the developing world might reveal interesting geopolitical relationships. Although humanitarian NGOs tend to move from crisis to crisis, those concerned with development and environment typically show regional preferences and are likely to establish deeper local roots. For example, North American environmental groups are more likely to work in Latin America. Their European counterparts are more likely to support projects in Africa.

More telling than mere numbers would be a study of the amount of capital NGOs transfer. NGOs, though, are private organizations and thus such financial data are difficult to obtain. One study estimates that northern NGOs sent $6.4 billion to the developing world in 1989, which is more than multilateral lending agencies lent that year.[9] Money aside, in the fields of environment and human rights, relatively small organizations have been key players in shaping global attitudes and policies. Understanding the spatial patterning of such NGO engagement and its impact requires more political geographic research.

Now that NGOs are experienced political actors, they are also receiving stiffer criticism. For decades, their flexibility, resourcefulness, innovation, cost-effectiveness, and participatory nature were praised as the antidote to ineffective public spending. Critics of NGO growth offer disparate reproaches. Some fault NGOs for being too small, limited in their reach, preoccupied with local issues to the neglect of national ones, dependent upon donations (often foreign), and unable to replicate successful programs. In sum, they may do good work locally but are unable to implement long-term structural reforms. At the other extreme, NGOs are criticized for becoming too big (especially in the North) with large professional staffs that are unaccountable, bureaucratic, and more concerned with their own institutional survival than with bringing about change. Both large and small organizations are challenged as to how representative and inclusive they are in regard to membership and clientele. All agree that the precarious financial base of NGOs is their greatest weakness. Given their nonprofit status, their existence depends on national and international donations, grants,

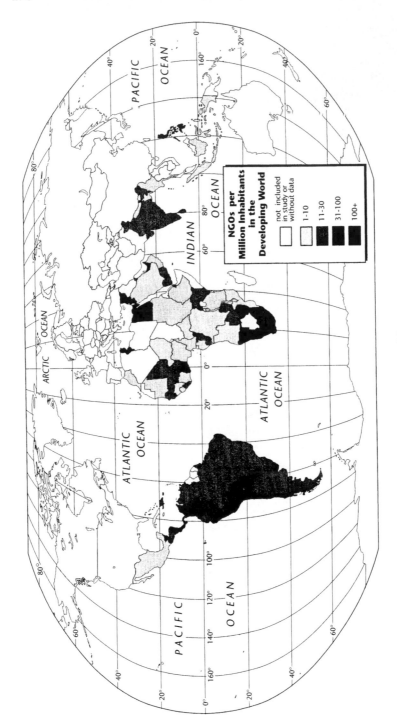

FIGURE 14.2 *NGOs per million inhabitants in the developing world.*

SOURCE: *Julie Fisher, Nongovernments: NGOs and the Political Development of the Third World (West Hartford, CT: Kumarian Press, 1997), 162–164.*

and government-funded work. Dependence upon foreign donations is tempting, but NGOs that hope to endure must develop sources of support other than state financing.

At the crux of the NGO critique is the question of public versus private provision of "public" services. NGOs are increasingly asked to provide traditional state functions: feed the poor, staff national parks, build housing, and improve water supplies. Many international donors and lenders see NGOs as an alternative to government ministries and are willing to fund their work. Most acknowledge that NGO services are important, but some fear that they should not be viewed as a substitute for sound government. Ideally, NGOs should work to improve government and intergovernmental policies, not replicate or replace them.

With the end of the Cold War a political aperture has opened for new geopolitical players other than the superpowers and their allies. NGOs stepped up their work in those areas where state concern was limited (environmental issues) or where complex humanitarian issues required neutral nonstate actors. Advocacy NGOs found that forceful lobbying and public education were effective tools in shaping policy, nationally and internationally. The World Bank's decision to include environmental considerations when making loans was a direct result of external NGO pressure along with internal institutional reform. The UN now considers NGO involvement essential to carrying out its many missions. And NGOs are welcomed at the conference tables of other intergovernmental organizations as well. At their best, these privately organized service-oriented groups are the last hope for politically and economically marginalized peoples. Even so, they are not a silver bullet to resolve poverty, political repression, and environmental degradation, despite the best intentions of NGO activists. Like all institutions, their integrity and effectiveness varies greatly. Their growing reach into the geopolitical arena should be of increasing interest to political geographers.

Notes

1. Several talented graduate students at the George Washington University have shared an interest with me in NGOs, and their work has informed my own. They include Rebecca J. Shumaker, Gillian Craig, Eloisa Moreas, David Cochran, Elizabeth Fraser, Ralph Espach, Kenn W. Rapp, and Blake Smith.

2. Jody Williams, the co-director of the International Campaign to Ban Landmines (an NGO) received the Nobel Peace Prize for her efforts in 1997. Despite the success of this organization, the vast majority of NGOs are little known outside the individual neighborhoods, communities, or regions they serve.

3. One can easily become lost in acronyms. Some of the more commonly used subcategories of NGOs include PVOs (private voluntary organizations, usually associated with developed countries), GROs (grassroots organizations), GRSOs (grass-

roots support organizations, larger groups that help finance GROs), QUANGOs (quasi-NGOs, or publicly sponsored NGOs that are organizational affiliates of a government ministry), CBOs (community-based organizations, much like GROs), NNGOs (northern NGOs based in developed countries), and SNGOs (southern NGOs based in developing countries).

4. Julie Fisher, *Nongovernments: NGOs and the Political Development of the Third World* (West Hartford, CT: Kumarian Press, 1997), pp. 6–7.

5. Fisher, *Nongovernments*, p. 52.

6. Ibrahim F. I. Shihata, "The World Bank and Non-Governmental Organizations," *Cornell International Law Journal* 25, no. 3 (1992): 623–641.

7. Thomas Princen and Matthias Finger, *Environmental NGOs in World Politics: Linking the Local and the Global* (London and New York: Routledge, 1994), pp. 3–4.

8. Estimates of NGO numbers are notoriously controversial. These figures rely on conservative figures offered by Julie Fisher in *Nongovernments: NGOs and the Political Development of the Third World*. They include local NGOs as well as GRSOs working in particular countries. Although the number of organizations is an indicator of NGO popularity, it tells us nothing about the size, economic influence, or political impact of these institutions.

9. John Clark, *Democratizing Development: The Role of Voluntary Organizations* (West Hartford, CT: Kumarian, 1991), p. 39.

References

Clark, John. *Democratizing Development: The Role of Voluntary Organizations.* West Hartford, CT: Kumarian Press, 1991.

Farrington, John, and Anthony Bebbington. *Reluctant Partners? Non-governmental Organizations, the State and Sustainable Agricultural Development.* New York: Routledge, 1993.

Fisher, Julie. *Nongovernments: NGOs and the Political Development of the Third World.* West Hartford, CT: Kumarian Press, 1997.

Korten, David C. "Third Generation NGO Strategies: A Key to People-Centered Development." *World Development* 15 (Supplement) (1987): 145–159.

Livernash, Robert. "The Growing Influence of NGOs in the Developing World." *Environment* 34, no. 5 (1992): 12–20, 41–43.

Meyer, Carrie. "The Political Economy of NGOs and Information Sharing." *World Development* 25, no. 7 (1997): 1127–1140.

Price, Marie. "Ecopolitics and Environmental Nongovernmental Organizations in Latin America." *Geographical Review* 84, no. 1 (1994): 42–58.

Princen, Thomas, and Matthias Finger. *Environmental NGOs in World Politics: Linking the Local and the Global.* London and New York: Routledge, 1994.

Shihata, Ibrahim F. I. "The World Bank and Non-Governmental Organizations." *Cornell International Law Journal* 25, no. 3 (1992): 623–641.

Wood, William B. "From Humanitarian Relief to Humanitarian Intervention: Victims, Interveners, and Pillars." *Political Geography* 15, no. 8 (1996): 671–695.

Chapter Fifteen

Global Hegemony Versus National Economy: The United States in the New World Economy

JOHN AGNEW

It was 250 years ago that the Scottish philosopher David Hume wrote of "the narrow malignity and envy of nations, which can never bear to see their neighbours thriving, but continually repine at any new efforts toward industry made by any other nation." Although this description rings true for much of modern history, if Hume's world of envious nations had run its course we would not have the interdependent world economy that is rapidly emerging today. But it did not. Beginning in the late 1940s and under U.S. government sponsorship, a largely successful attempt has been made to reconcile the contradiction between world capitalism and world politics by creating a global economy in which production and finance increasingly operate at a global scale. The vision of a world that knows no boundaries is at odds with a world of independent states jockeying for power and advantage. It has also had paradoxical effects on the position of the United States in the contemporary world. Although the United States is acquiring an apparent military and cultural hegemony at the global scale, the American vision of a world economy that transcends national differences has suffered because of blows to its own national economy. This has led to a loss of faith in the "American Dream" of continued upward social mobility and increasing consumption as income inequality increases in general and between regions and within metropolitan areas. The outlook is for growing social polarization and political disaffection.

The Cold War is won, positive images of American and Western European market capitalism sweep Eastern Europe and the former Soviet Union, the world's largest McDonald's restaurant has opened in Beijing, and Disney has a theme park in France. The American economy has achieved much lower levels of unemployment than those found in most other market economies. The New York Stock Exchange has swept to highs in average gains-per-share traded without parallel in modern history. At this historical moment of cultural and economic dominance, however, a wave of self-doubt sweeps across the United States. Even as unemployment drops to relatively low levels compared to the late 1980s and early 1990s, opinion polls suggest that two-thirds of Americans believe their country is "on the wrong track." For the first time in more than a generation, parents do not believe that their children will have a better life than they themselves have had.[1]

Many economists do not share this pessimism. They contend that the American economy is ripe for a long-term boom. They point to economic indicators: Employment is up, growth is steady, and inflation is down. Beyond the statistics, they suggest that firms have aggressively cut costs and improved profitability, Congress has eased regulation in key industries, such as banking and finance, and new technologies, particularly in computing and information processing, promise improvements in productivity and work flexibility. Nevertheless, the job security and relatively high wages in manufacturing jobs that characterized the American economy as recently as twenty years ago are seen as things of the past. To be competitive in the new world of global competition, with diminished trade barriers and massive flows of cross-border investment, firms must be "lean and mean." American firms can no longer worry that their decisions about where and how much to invest will have disproportionate impacts on Americans. In a world that knows no borders they can no longer afford the costs of nationality.

In this chapter I argue that the very success of U.S. global hegemony has imposed costs on the American national economy, produced local economic problems, and generated the national mood of pessimism. The New World Order made under American auspices is a "transnational liberal economic order" that increasingly detracts from, rather than augments, the material conditions of life for both the "average" and the poorest people within the territorial boundaries of the United States.[2] The economic impacts of America's global hegemony also vary from place to place in the United States. This chapter surveys the evidence for this claim, lays out a suggested explanation for how U.S. influence has produced a "globalized" world economic order, and identifies specific economic processes producing the negative consequences within the U.S. "territorial economy."

Postprosperity in the United States

Income Divergence

The boom years for the post–World War II American economy ended in 1973. In that year, the median earnings for males peaked, then entered a twenty-year period of decline and stagnation. Median household incomes have also stagnated and have only held their own because of increases in paid employment for women. During the same period, average family income did inch upward—a rate of increase of 0.04 percent per year, compared to the 2.72 average of the fifteen years prior to 1973—but this masks a significant redistribution of income within the population. If the distribution of family incomes is divided in tenths (deciles) and if 1977 is compared to 1988, it can be shown that real (after-inflation) incomes declined for every tenth except the highest two. In the lowest tenth, family incomes (which averaged an amazingly low $3,504 in 1988) were 14.8 percent lower in 1988 than in 1977 in constant dollars. In the middle-income groups—the fifth, sixth, and seventh tenths—average family incomes were lower by 5 percent. Even the eighth tenth saw a shrinkage in incomes, off 1.8 percent in constant dollars. The ninth decile had a minor increase of 1.0 percent; the biggest increase, 6.5 percent, occurred in the wealthiest tenth decile, which captured most of the increase in gross national product (GNP) over the ten-year period.[3]

These data suggest a decline in prosperity for most Americans after a period in which improvement had become an almost universal expectation. Perhaps more significant, the income distribution has also become more polarized. More Americans are rich now, and more are poor. Fewer have stayed in the middle-income groups. In the 1980s, the "middle class," which has given the United States its reputation for high mass consumption and political stability, shrank by about 10 percent. One study estimates that in the 1980s, 2 percent of those in the $18,500–$55,000 (1987 dollars) income group (75 percent of the population in 1980) rose into the upper income brackets and that 30 percent slid downward.[4] With fewer people moving into the middle range from either end of the distribution, the net result is the politically controversial phenomenon of the "shrinking" middle class.

The most proximate cause of the upward redistribution of income and the erosion of the middle class is the changing structure of the job market. High-skilled workers are earning more and lower-skilled workers are earning less as the American economy uses fewer highly paid, unskilled laborers than it did during the "glory days" after World War II. The earnings gap between the rich and poor began to widen in the late 1960s but increased even faster in the 1970s and 1980s. To be sure, income polarization is occurring in other industrial economies, but elsewhere, especially in Europe, family earnings have been more protected by government benefits, and fewer people have fallen into the low-income groups.

Behind the change in the job market lies a fundamental restructuring of the American economy in the face of increased exposure and vulnerability to the world economy. More simply put, the increase in U.S. trade and foreign investment over the past twenty years bears much of the blame for the surge in income inequality and the shrinking size of the American middle class. From 1979 until 1989, as the sum of exported and imported goods rose from 55 percent to 82 percent of U.S. manufacturing output, the average pay and benefits of American factory workers fell 6 percent after inflation, even as plant productivity rose 42 percent.[5] At the same time, job growth in the American economy has been concentrated in the services sector, which is strongly polarized between low-earning and high-earning jobs. Manufacturing jobs traditionally provided America's middle incomes. In 1975, 45 percent of manufacturing jobs had incomes that clustered close to the national median. In contrast, the incomes of service-sector jobs are skewed downward or are distributed bimodally; for example, those in the retail sector (a major growth area in the 1980s and early 1990s) are skewed downward, and the producer-services sector is bimodal. Increased openness of trade in services promises some new markets for American firms, but it also makes service-sector jobs subject to foreign competition in the same way that manufacturing jobs have been for much longer.[6]

Regional and Intrametropolitan Divergence

New low-wage service jobs do not compensate adequately for the lost manufacturing jobs and lower incomes in manufacturing brought about by the American economy's increased involvement in world trade.[7] The widening inequality of incomes this creates poses major challenges to the country. As poverty among people in the peak years of employment increases, the costs of welfare and unemployment will mount, and resentment of the wealthy may well grow. One new trend is that, after thirty years of convergence, average regional per capita incomes began to diverge in the 1980s.[8] This reverses the nation-building trend of the postwar period and opens the long-term possibility for regional fragmentation like that experienced in the aftermath of the Cold War in Eastern Europe and the former Soviet Union. Much of the rural U.S. South remains poor, with inadequate school systems and high rates of illiteracy, and all regional and metropolitan economies show startling variations in incomes and prosperity. Regions with large minority populations display particular signs of decline in these areas. As income inequality increases, so does ethnic and social-class segregation.[9]

Not all of this geographical variation is due to the direct effects of trade on the manufacturing industry. Fluctuations in defense spending, cutbacks in government welfare programs, and the concentration of the most dynamic high technology manufacturing (especially computers) and high-growth service (especially finance) industries have all played their part. Cutbacks in defense spending have particularly affected those regions, such as New England and southern California, that had become especially dependent on the militarization of the U.S. economy during the Cold War.[10] The impact is clear: As U.S. manufacturing trade grew by 60 percent in the 1972–1985 period, the average annual income of 17 million professionals was 9 percent higher than it would have been otherwise, and the average annual income of 90 million other workers was 3 percent lower than it would have been without the growth in trade.[11]

What happened? From one point of view, the period after World War II was exceptional. The U.S. economy reigned supreme in a world not yet recovered from the mass destruction of war. After the depression and war, there was an unprecedented potential for capital investment (at home and abroad) and for domestic consumer spending. Federal government activism in the form of the interstate highway system and civil rights legislation integrated the South, previously a backwater region, into the national economy. National housing legislation and highway development allowed the massive suburbanization that stimulated a large number of industries. Meanwhile, industrial production was largely organized around national markets with limited foreign competition. The *Pax Americana* of the postwar years gave the U.S. economy significant leverage over the world's financial and trading systems, even though the United States itself was only modestly exposed to the vagaries of either. But after 1973, this exceptional political economy ran out of steam.

Two events of 1972–1973—the U.S. government's abrogation of the Bretton Woods agreement that had stabilized the world monetary system after World War II and the OPEC oil price increases—jointly symbolize the emergence of a world economy to which the U.S. national economy was increasingly exposed and vulnerable. If the world economy seemed a benign or beneficent influence on the American economy before 1973, everything changed thereafter. Eurocurrency markets (using U.S. dollars in circulation outside the United States) became more important, and international banks began to privatize and globalize the international financial system, thereby effectively reducing U.S. government control over capital. The growth of producer cartels, especially OPEC, and of the European Community in the 1960s limited U.S. control of vital raw materials and foreign markets. And mounting trade deficits signaled the arrival of a more internationalized production, from which the United States could not be sheltered without imposing costs on its own businesses. At the same time, military-political reverses in Vietnam (and later in Iran) challenged American preeminence within its various alliances and undermined the political consensus in the

United States that sustained both the U.S. military presence overseas and the growing welfare state at home.[12]

The United States and Globalization

To portray the U.S. territorial economy as a "victim" of the worldwide growth in trade and investment would be misleading in two respects. First, many Americans and some specific American localities have benefited from increased trade and foreign investment. This is particularly true for employees and places associated with successful export firms (such as Boeing and Microsoft in Tacoma-Seattle) and those benefiting from the inflow of foreign direct investment (the Toyota and Honda factories in, respectively, Kentucky and Ohio). Second, and more fundamentally, the contemporary world economy is largely a product of U.S. design and ideology. Since the late nineteenth century, American business and political leaders have been major sponsors of an integrated and interdependent world economy. The U.S. economy was the first industrial economy to abandon a territorial mode of operation for what can be called an *interactional mode of operation*. After the U.S. continental frontier closed in the 1890s, American businessmen focused on commerce and trade largely independent of the territorial entanglements that constrained and drained their counterparts in Europe and Japan. Paralleling this was a progressive expansion in the geographical scope of U.S. economic interests—from continental and hemispheric scales of emphasis in the early 1900s to a global scale after World War II.

At the root of the American rise to hegemony and the creation of the world economy as we know it today lay two features of the U.S. historical experience. First, America's own colonial past made territorial colonialism an ideologically difficult enterprise; U.S. institutions claimed their origins in colonial revolt rather than dynastic or national continuity. Second, after the Civil War, an integrated national economy emerged that was increasingly dominated by large firms; as they developed overseas interests, these firms were able to shape the American international agenda. For many years, the division of the world into trading blocs and territorial empires limited U.S. influence. Powerful strains of public opinion were also opposed to American involvement in foreign economic and political affairs. After World War II, however, an intensely internationalist American agenda, sponsoring free trade, currency convertibility, and international investment, was advanced in explicit counterpoint to the autarkic dogmas of Soviet communism and as a response to the competitive trading blocs that were seen as partly responsible for the depression of the 1930s. The effort to design a "free world" order in the immediate postwar years laid the groundwork for the internationalization of economic activities in the 1960s that brought tremendous expansion in U.S. firms' investment overseas and the increased importance of trade for the U.S. territorial economy.[13]

Between 1960 and 1970, new American corporate investments abroad expanded from 21 percent to 40 percent of total investments. Much of this reflected the desire to augment markets by avoiding trade barriers, such as tariffs and quotas. But some of it involved a search for lower wage costs on the part of labor-intensive industries. For example, Motorola now has only 44 percent of its 100,000 employees in the United States versus nearly 100 percent in 1960. At the same time, the products of Motorola and of other firms, both American and foreign, were building markets in the United States as imports. A "global shift," or internationalization of production, was under way.[14]

The new transnational economic order has three important features that set it apart from those of earlier periods. First, foreign direct investment among major industrial countries has increased at a faster rate than has the growth of exports among them. The ties that bind industrialized economies together are those of investment more than trade. In the 1980s and early 1990s, the rate of growth of foreign direct investment in the world economy has been three times that of the growth of world exports of goods and services.[15]

Second, national trade accounts can be misleading guides to the complex patterns of trade and investment that characterize the new global economy. Perhaps 40 percent of total world trade between countries as of 1990 was trade within firms.[16] Further, more than half of all trade between the major industrial countries is trade between firms and their foreign affiliates. A third of U.S. exports go to American-owned firms abroad; another third goes from foreign firms in America to their home countries. And because the new global trading networks involve the exchange of services as much as the movement of components and finished goods, many products no longer have distinctive national identities. The U.S. 1986 trade deficit of $144 billion thus becomes a trade surplus of $77 billion if the activities of U.S.-owned firms outside the United States and foreign-owned firms in the United States are included in the calculations.[17]

Third, as the U.S. territorial economy loses manufacturing jobs and shares of world production to other places, the global shares of its firms are maintained or enhanced. As the U.S. share of world manufactured exports went from 17.5 percent in 1966 to 14.0 percent in 1984, American firms and their affiliates increased their shares from 17.7 percent to 18.1 percent.[18] This leads to the question, "Who is Us?" in relation to government policies that can favor U.S. firms rather than the U.S. economy.[19] From this point of view, helping "foreign" firms locate in the United States benefits the U.S. territorial economy more than helping "American" firms, which may be owned by Americans or headquartered in the United States but have most of their facilities and employees located overseas.[20]

Economic Mechanisms and Negative Consequences

How has the "globalization" of the U.S. economy produced a redistribution of incomes and a shrinking middle class? Three interlocking mechanisms have been particularly important in producing these negative consequences. First, U.S. government macroeconomic policies and the activities of American firms have increased the exposure and vulnerability of the U.S. economy to international competition. The General Agreement on Tariffs and Trade (GATT) and bilateral agreements on limiting impediments to trade began to open up the U.S. economy in the 1960s, while American firms began globalizing their operations to serve foreign markets and substitute foreign for domestic production in serving American ones.[21] The net effect has been a reduction in the U.S. share of total world production (although the United States remains the single largest producer) and an increasing American involvement in world trade (as measured by U.S. exports as a share of total world exports). Consequently, as goods have been traded more freely, prices and production costs have been set globally. Labor-intensive goods are more cheaply produced in low-wage countries, so low-skilled work formerly done by American labor (sometimes under generous contracts negotiated by trade unions) has flowed overseas. Competitive trading has also led to declines in the incomes of low-skilled U.S. workers. In addition, the recent political weakness of labor unions in the United States has meant that American firms have found it relatively easy to adjust to changed conditions at the expense of low-skilled workers.[22]

Second, the U.S. economy suffers from a relatively low rate of fixed investment in manufacturing plants, research and development, and employee training compared to most other industrialized countries. The consequences of this are lower productivity increases relative to many other countries, a drop in national economic competitiveness (particularly relative to Japan and Germany), and a frantic search for quick fixes to counter declines in profitability. Much-publicized fixes have included payoffs, employee "paybacks," moving to states or municipalities that offer fiscal "sweeteners" such as tax breaks, moving offshore, stripping assets from established operations, and moving into new sectors.

Many factors have been at work in "hollowing out" the American economy. One is the increasingly fragmented and transient nature of firm ownership in the United States.[23] Institutional investors, such as pension funds, have increased their share of total firm equity from 8 percent in 1950 to 60 percent in 1990. This gives managers considerable autonomy but at the expense of long-term investment. For example, in the 1980s, many managers and "corporate raiders" preferred the short-term results coming from mergers and acquisitions to such long-term strategies as investment in new products and worker training. Of course, one effect of this in the 1980s was to increase employment and incomes in the financial services industry. Another factor accounting for low investment is America's low savings rate. Funding personal consumption has increasingly taken priority over saving and investment, which has affected both business fixed investment and public investment in infrastructure. The lack of investment in physical infrastructure (roads, sewers, and so on) may have had a particularly negative impact on productivity since the late 1960s.[24] Although capital markets are increasingly globalized and American firms can draw on foreign sources, there is still a fairly high positive correlation between domestic savings and domestic investment across a range of countries, including the United States.[25] By 1990, 58 percent of national savings were absorbed in paying the U.S. federal budget deficit, compared to the 2 percent average of the 1960s. Thus, the federal deficit now operates as a drag on the national economy by absorbing domestic savings and by diverting potentially productive foreign portfolio investment into debt service.

A large portion of the U.S. current account deficit (the gap between what the U.S. territorial economy "earns" outside and the total amount that leaves the United States) is attributable to foreign borrowing to cover the federal budget deficit (averaging between $400 and $250 billion in the years 1992 to 1996, or between 6.7 and 3.8 percent of GNP). This cuts into American living standards by sending abroad payments that could be invested in the productive capacity of the U.S. territorial economy. The debt's effect on interest rates and the value of the U.S. dollar against other currencies also constrains the U.S. government's use of fiscal and monetary policy to manage the economy. Since the late 1970s, the American economy has become trapped in a vortex of currency and interest-rate volatility as successive governments have tried, through accords with foreign governments and shifts in interest rates, to create a better national economic climate.[26] The dollar, as the major standard for global exchange, remains a major weapon in the hands of U.S. policymakers now that the federal deficit has reduced the possibility of government spending being used as an economic catalyst. American leverage over the world economy and domestic investment has faded as a result of increased openness in that world economy and low levels of private investment; it has also been squandered by successive national governments that increased military and entitlement (e.g., social security) spending without raising taxes.[27]

The third "mechanism" that is squeezing the middle class is the decline in American technological leadership. The relative competitiveness of industries depends, in part, on their ability to identify and exploit new products and production technologies. In the aftermath of World War II, the United States had no peer in this regard. Since the 1970s, however, the U.S. patent balance (a measure of the flow of innovative technologies) with Japan and Germany has turned increasingly negative. Although U.S. firms now find themselves faced with entire markets dominated by foreign producers (such as that for semiconductors), they apparently continue to give relatively low priority to investment in research and development (R&D). More importantly, the huge U.S. government investment in science and technology over the past few decades has largely been in areas connected to military applications rather than market-oriented research. With U.S. R&D dominated by the defense sector, foreign competitors have been able to go "straight to market" with innovations; meanwhile, the American technology policy required slow and costly justifications in terms of national security.

The end of the Cold War may allow the U.S. government to reorder its technological priorities. Even so, technologies and new innovations are the property of the firms that bought or created them, rather than the territories in which they were developed. With economic globalization, technologies are no longer tied to particular national economic spaces or to increasing the productivity of a national or local economy. The incomes of those who invent and develop technologies can increase without a "filtering down" of benefits to workers and communities. And even high-end R&D activities can also move abroad, as has happened with IBM and Motorola.

Increased openness places a premium on the ability of a territorial economy to generate the investment and technologies that can meet the challenge of foreign competition. Failure to do so in a "free trade" world exacts a heavy price for a national economy. But American firms regard the world as their oyster; they are not necessarily constrained by loyalty to the U.S. territorial economy if they can be more successful elsewhere. As one businessman declared, "The United States does not have an automatic call on our resources. There is no mindset that puts the country first."[28] This is at the heart of the seeming paradox between American global hegemony and the changed condition of the U.S. territorial economy.

Conclusion

There is nothing "natural" or inevitable about a world divided into national or territorial economies. Under American influence, the growth in trade and financial flows has produced a world economy in which even U.S. hegemony cannot guarantee a perpetual dividend for America's national economy. For many Americans, there is a divorce rather than a marriage of common interests between the U.S. and world economies. The "free world" that the Cold War was fought to sustain no longer seems so unambiguously a good thing. After the Cold War, Americans are not the only ones asking the "who is us?" question. The consequences of increased trade and capital mobility for all national economies and the incomes of relatively immobile populations everywhere are beginning to excite tremendous popular interest all over the world. Yet discussions about geopolitics and economic development continue to be cast largely in terms of the national territories and "national interests" that David Hume identified in the mid-1700s. It is as if the world economy as we know it today did not exist. Superficially, the United States now stands in the position it occupied in 1945. It was the world's only military superpower then, and so it is now. It was the world's largest economy then, and so it is now. And if anything, its cultural influence is greater now than then. But Americans are no longer confident of its position. Like "ordinary" countries, the United States now feels the pressure of the outside world on its economy, and many Americans fear internal social fragmentation will result. The new world economy has no natural sympathies, even in the homeland of those who did the most to create it.

Notes

Earlier versions of the first edition of this chapter were presented as lectures at the University of Macerata, Italy (June 10, 1991), and at Cambridge University, England (February 9, 1992). The interesting questions and suggestions that I received in both places have helped me in writing the chapter for this volume.

1. W. Wolman and A. Colamosca, *The Judas Economy: The Triumph of Capital and the Betrayal of Work* (New York: Addison-Wesley, 1997).

2. S. Gill, *American Hegemony and the Trilateral Commission* (Cambridge: Cambridge University Press, 1990).

3. W. Peterson, "The Silent Depression," *Challenge* (July-August 1991): 30–31; K. Phillips, *The Politics of Rich and Poor* (New York: Random House, 1991).

4. G. J. Duncan, T. M. Smeeding, and W. Rodgers, *W(h)ither the Middle Class? A Dynamic View,* Income Security Policy Studies Series, no. 1 (Syracuse, N.Y.: Metropolitan Studies Program, Maxwell School, Syracuse University, 1992).

5. M. H. Kosters, *Workers and Their Wages* (Washington, D.C.: American Enterprise Institute, 1992).

6. R. Erzan and A. J. Yeats, "Implications of Current Factor Proportions Indices for the Competitive Position of the U.S. Manufacturing and Service Industries in the Year 2000," *Journal of Business* 64 (1991): 229–254.

7. Y. K. Kodrzycki, "Labor Markets and Earnings Inequality: A Status Report," *New England Economic Review* (May-June 1996): 11–25.

8. See Phillips, *The Politics of Rich and Poor.*

9. J. E. Kodras, "The Changing Map of American Poverty in an Era of Economic Restructuring and Political Realignment," *Economic Geography* 72 (1997): 67–93; C. J. Mayer, "Does Location Matter?" *New England Economic Review* (May-June 1996): 26–40.

10. Office of Technology Assessment (OTA), *After the Cold War: Living with Lower Defense Spending* (Washington, D.C.: OTA, 1992).

11. E. E. Leamer, *Wage Effects of a U.S.-Mexican Free Trade Agreement*, Working Paper 3991 (Cambridge, Mass.: National Bureau of Economic Research, 1992).

12. J. A. Agnew, *The United States in the World Economy: A Regional Geography* (Cambridge: Cambridge University Press, 1987).

13. J. A. Agnew and S. Corbridge, *Mastering Space: Hegemony, Territory, and International Political Economy* (London: Routledge, 1995).

14. P. Dicken, *Global Shift: The Internationalization of Economic Activity* (New York: Guilford Press, 1992).

15. "Globetrotting," *Economist*, July 27, 1991, p. 58.

16. R. B. Reich, *The Work of Nations: Preparing Ourselves for 21st-Century Capitalism* (New York: Knopf, 1991).

17. D. Julius, *Global Companies and Public Policy: The Growing Challenge of Foreign Direct Investment* (New York: Council on Foreign Relations, 1990).

18. R. Lipsey and I. Kravis, "The Competitiveness and Comparative Advantage of U.S. Multinationals, 1957–1984," *Banca Nazionale del Lavoro Quarterly Review* 161 (1987): 81–96.

19. R. B. Reich, "Who Is Us?" *Harvard Business Review* (January-February 1990): 53–64.

20. L. D. Tyson, "They Are Not Us: Why American Ownership Still Matters," *American Prospect* (Winter 1991): 37–49; R. B. Reich, "Who Do We Think They Are?" *American Prospect* (Winter 1991): 49–53.

21. "American Firms Have Tended to Have More Interest in Foreign Markets Than in Producing for Sale in the U.S.: Balancing Act," *Economist*, January 4, 1997, p. 71.

22. D. Rodrik, *Has Globalization Gone Too Far?* (Washington, D.C.: Institute for International Economics, 1997).

23. M. Porter, *Capital Choices: Changing the Way America Invests in Industry* (Cambridge: Harvard Business School, 1992).

24. A. H. Munnell, "Why Has Productivity Growth Declined? Productivity and Public Investment," *New England Economic Review* (January-February 1990): 3–22.

25. T. Bayoumi, *Saving-Investment Correlations*, Working Paper 89/66 (Washington, D.C.: IMF, 1989).

26. R. McCulloch, "Macroeconomic Policy, Trade, and the Dollar," in *The Economics of the Dollar Cycle*, ed. S. Gerlach and P. A. Petri (Cambridge: MIT Press, 1990).

27. B. M. Friedman, *Day of Reckoning: The Consequences of American Economic Policy* (New York: Random House, 1989); D. P. Calleo, *The Bankrupting of America: How the Federal Budget Is Impoverishing the Nation* (New York: Morrow, 1992).

28. L. Uchitelle, "U.S. Businesses Loosen Link to Mother Country," *New York Times*, May 21, 1989.

References

Agnew, J. A., and S. Corbridge. *Mastering Space: Hegemony, Territory, and International Political Economy.* London: Routledge, 1995.

"Earnings Inequality: Proceedings of a Symposium on Spatial and Labor Market Contributions to Earnings Inequality," *New England Economic Review,* Special Issue (May-June 1996).

Rodrik, D. *Has Globalization Gone Too Far?* Washington D.C.: Institute for International Economics, 1997.

Wolman, W., and A. Colamosca. *The Judas Economy: The Triumph of Capital and the Betrayal of Work.* New York: Addison-Wesley, 1997.

Chapter Sixteen

Geopolitical Information and Communications in the Twenty-First Century

STANLEY D. BRUNN, JEFFREY A. JONES,
AND SHANNON O'LEAR

Information technologies and communications are an often overlooked area of investigation in political geography. Geopolitics is centered on the state, but the very definitions of *state* and *nation* are challenged by worlds made possible through advances in information and communication technologies. Some immediate questions for political geographers who examine the role of communications are, What are the key components in political information flows and how have these flows changed over time in different societies? Who uses these information flows and what type of landscape is constructed as "reality"? How is the political landscape constructed and maintained by the use of certain information and the suppression of other points of view? This chapter sketches an exploratory model for examining these issues and considers the role of information and communication in future political geographical analysis. Among the themes we examine are labels and symbols, worldviews, the "electronic state," and the changing nature of geopolitical information in 1900, 2000, and 2100.

The Role of Information

Consider the influence of photographs and film footage on public opinion and resulting political actions of these events during 1997:

- China assuming control of Hong Kong with festive celebrations;
- Kabila marching toward Kinshasa to assume control of the Zaire government from Mobutu;
- the state funerals of Princess Diana and Mother Teresa;
- government representatives signing the global climate treaty in Kyoto;
- Swiss officials responding to charges that banks hoarded the gold of Holocaust victims;
- the collapse of the Asian stock markets;
- testimony of wrongdoers before South Africa's Truth and Reconciliation Commission; and
- the Pathfinder rover exploring the surface of Mars.

A few years earlier powerful televised photographs showed starving children in Somalia and crying children, emaciated adults, and burned corpses in Bosnia. And prior to that were the first photographs of planet Earth taken from a spaceship, a "blue ball" against the darkness of space. We usually gauge our reactions to a significant event—the assassination of John F. Kennedy, the explosion of the space shuttle *Challenger,* the Tiananmen Square massacre in June 1989, and fall of the Berlin Wall—not by the time of the event but by the time and place we first learned of it. Whether "being there" is a newsreel of the Kennedy assassination made before one's birth or the mere seconds of delay in live televised reports, the medium and content of information define these events and places.

The power of communications to affect our construction of reality and truth raises a very important question: How do we interpret events via an electronic medium? Infodramas and docudramas portray historical figures and social analogies according to certain slants to impress on the viewer a certain message or to enhance entertainment appeal. But just how "real" or objective are these characters and situations to viewers? Is the Columbus who was celebrated in 1492 as a hero any less historically correct than the Columbus of 1992 who was reviewed in society and portrayed in the media as a conveyor of genocide and environmental destruction? Future geographic inquiry will likely focus on how people incorporate these images of events, people, and places as "real" versus entertainment, as one possible interpretation versus "the truth." The state will be one of the voices heard and seen, but it will vie with other producers, including other politically motivated groups and entertainment producers, to provide facts, interpretations, and "truth." Information and images have had direct connections to subsequent political inquiry in all types of countries and at all levels, local to global. It is this interconnectivity across scales and between information and policy that interests political geographers.

Information in a Geopolitical World

The progress or failure of any political unit is dependent on a series of information-related questions starting with David Smith's definition of welfare geography, Who gets what from whom where? In the context of the political geographies of information, the *who* is usually the state, the *what* is all varieties of information, the *from whom* is the source or producer of the information, and the *where* is the location of what is produced, consumed, and exchanged.

Information is a commodity that can be used for beneficial or harmful purposes. The information produced is often a symbol of power, be it a state, a corporation, or a special interest group. Information "goods," such as a textbook, a map, a television or radio program, a law, or an official photograph of an event, merit scrutiny for their impacts on peoples and places. Recent political geographic studies have examined such information-related phenomena as the siting of noxious facilities and the delivery of public services (health, education, and welfare goods). Likewise, power, ideology, legislation, and constitutions are all linked to information; all are influenced by what information is available, how it is used or misused, and how it affects people and place.

The State as Text and Discourse

Geographers have begun to look at landscapes as text: products "written" by certain "authors" to reflect a specific representation and voice. Indeed, the state can be seen as a text produced to reflect a certain national image and redrawn through the political process. Tied to this dynamic process, however, is the concept of "discourse." French philosopher Michel Foucault defined a *discourse* as interplay of power relationships in which knowledge is constructed within a set of sociopolitical rules. Language, concepts of science and society, and political goals are components of a discourse. Since the Enlightenment at least, Western science has fostered a dominant discourse based on repeatable observations in line with a "scientific" methodology. This discourse places experiences seen or felt only by a single person and processes not readily repeatable, such as religious or extrasensory experiences, outside of scholarly legitimacy. Language is another discourse, with its own nuances and values. Language becomes an integral part of identity and interpretation. American history reveals the dynamic changes in the discourse producing the political landscape. Tied to this changing discourse are questions of *who* speaks in the discourse and *what* structures are maintained by it. Designed and implemented largely by wealthy colonial men of European heritage, the American system of government legitimizes and creates a formal discourse of elections and representative democracy. The U.S. Constitution scripts a political landscape, and the electoral political process and government system thus created provides the set of rules for what constitutes a legitimate political voice in the United States. This single document defined the political geographies of the United States.

Before 1900 the majority of Americans were not legally allowed to express political concerns through the discourse initiated by the Founding Fathers; the legitimized political landscape was defined officially by the voice of white, adult males. Alternative voices had to be expressed in other ways, for example, through fiction, art, music, and the popular press. Denied political expression, Harriet Beecher Stowe influenced the politics of her day via her novel *Uncle Tom's Cabin*. Similarly, African Americans and others used music and the popular press. Legitimacy and enfranchisement are granted by those already holding power. Changes in the political landscape can thus be viewed as the result of pressure by various groups outside the legitimized discourse to influence the dominant speakers via alternative information channels.

Nationhood is also a social construction with an often vague rationale. Benedict Anderson has called nations "imagined communities" because they rely on a common concept of unity that often defies differences. Most Americans will never meet other Americans from other regions, and probably New Englanders share more culturally with Canadians than with people in Hawaii, and many Texans have a history and language more linked to Mexico than to the United States. Yet most people in New England, Hawaii, and Texas view themselves as "Americans." The highest percentages of those defining themselves as Americans, that is, not with any ethnic or racial label, occur in the rural South. How has this commonality of American nationality come into being? The state serves as a partner in the creation of a sense of nationhood, and states benefit directly from those unifying practices.

Anderson also points to the role of what he calls the creole class in developing nations in the Americas. Creoles, or people of European parentage born in the colonies, were historically the dominant class in colonial social hierarchies. Creole youth educated in the home country often found a common identity with other creoles, who were familiar with their foods and customs. Creoles sometimes found that the colonial politics dictated by the imperial government were unfavorable to maintaining their privileges and self-interests. What was good for the home country and its standards of morality was not necessarily good for the creole colonials. The revolutions for independence were often portrayed by the resulting states as populist revolts. Yet for many of the lower socioeconomic classes, such as slaves and native populations, colonial policies may actually have been more lenient or protective than the laws of the newly independent countries. Anderson sees the landed creole gentry as the driving force behind colonial revolutions; the subsequent scripting of popular opinion and history by the state fostered a sense of populist revolution and national commonality.

Labels: Who Is What?

We live in a world where labels are powerful, whether used by citizens to identify themselves or to distinguish them from "others" around them. Governments use labels to promote affinity with or antipathy against others. Strong ties between Europe and North America were important in collaborating on military and economic policies against declared enemies. European colonial powers referred to their commonwealth or community properties as extensions of themselves. States through their regulations and terminology officially define visitors, refugees, and documented and undocumented immigrants. Some African Americans align themselves with groups in Africa who are fighting for self-determination, just as some American Jews support militant policies of certain Israeli leaders. Labels can also convey misleading and inaccurate meanings, and the state can use them for propaganda purposes. The term "Islamic fundamentalism" is often a euphemism for militants, terrorists, and anti-U.S. governments. Both the Israelis and Palestinian leaders frequently use the words "terrorist" or "aggression" in official statements to justify their actions and to seek international support. Likewise, some religious groups may use the word "Christian" to garner support for specific leaders or proposed government initiatives.

When states define groups with labels, they convey identity. National identity cards are issued by many countries that utilize labels defined and accepted by the state. A very common labeling in the United States is *ethnicity,* and it has been institutionalized in decennial census counts. Those ethnicities so identified are categories or labels (or coded information) approved by the government. The U.S. Bureau of the Census recognizes four groups into which the citizenry is divided: white, black, Native American, and Asian or Pacific Islander. As the number of interracial marriages continues to climb, however, fewer and fewer people fit neatly into these categories. It was suggested to Congress in 1997 that a fifth category of "multiracial" be added, and seven states have since agreed to include a mixed-race category on school enrollment and other government forms.

Labels are the foundation of "ethnic cleansing." This practice was used by Hitler to persecute Jews and by Europeanized populations in North and South America to exterminate native populations or place them on reservations. During the recent war in former Yugoslavia, the Serbian army engaged in ethnic cleansing to eliminate Muslims from areas considered part of "greater Serbia." The Balkan peninsula is comprised of many Slavic cultures or groups, who through intermarriage over centuries are related. The three major groups in former Bosnia are the Bosniaks (Muslims), Croats (Catholics), and Serbs (Eastern Orthodox), but efforts to classify people on the basis of a single ethnic identity often are misleading. States adopt labels, define citizenship, and illustrate cultural data to serve specific political purposes.

Aside from the state denoting acceptable ethnic or racial labels for its residents, it makes decisions about others within its jurisdiction that relate to services provided residents and their political rights. Each state establishes the legal definition of what a citizen is and what noncitizens must conform to in order to become citizens; usually this requires a residence period, cash outlay, and official examination. States with more open borders and easier flows of short-term migrants use labels to carefully define qualifications for residence and citizenship. The status of noncitizens is a controversial political issue in many states, not only in regard to what label they should be assigned but also to what extent they are entitled to social services (health care, education for children, etc.) and political rights.

Symbols of the State

Another way that states send messages to their own citizens and other states is through a variety of symbols, including flags, logos, currency, and stamps. Colonial powers historically used commercial firms, military forces, missionaries, and scientific expeditions to convey power, identity, and legitimacy. They also utilized posters, maps, as well as holidays, theater, music, and school contests to promote knowledge about the country and its colonial possessions. Maps on stamps, for example, became a part of the information colonial powers used to show their place in the world or that of their possessions. With independence, stamps and currency became important vehicles for new states to show their status in the world. Likewise new emblems and institutions were shown on national currencies, and new monuments were erected as old ones were destroyed. Streets and public parks were renamed, and new flags flew over public buildings.

Several examples illustrate the impact of political symbols. Leaders of the Palestine Liberation Authority, for example, desired not only their own flag, constitution, and postage stamps in reclaiming possession of their territory but also wanted to establish their own national airline. New Central Asian states take pride in new historical displays of their history in state museums and in commemorating ethnic holidays and festivals banned during decades of Soviet rule. Nationalism is promoted through their own television and broadcasting programming, albeit often crude by European standards. The official state language is now that of the titular (or ethnic) majority and Russian has been reduced to a secondary status. Citizenship has been redefined.

Worldviews

Worldview is a state's perceptions of itself and of others. All states have perceptions of others. They may identify themselves as being in one or more categories, for example, with small or powerful states, as a traditional or emerging regional power, or one whose influence is waning because of major structural changes in the world. States also express their views of others by labeling them friendly or unfriendly, oppressive or democratic, aligned or neutral. A state's worldview can be seen in how it presents news to its citizens, how it represents others, what freedoms it grants to journalists, and even how political cartoonists depict political leaders and global problems.

How a state views itself and the world becomes increasingly important as images, symbols, and actions are observed by others. A state may seek to instill a worldview through the use of favorable propaganda disseminated to adults, small children, and others outside the state. Russia, Switzerland, Israel, Australia, and the United States, as well as Britain and other European colonial powers, developed and promoted images of themselves that instilled pride, patriotism, and positions of "specialness," often including divine providence and intervention. Powerful military and industrial states used commercial and religious communities to promote this position of "exceptionalism."

A state can express its worldview in a variety of ways. For its own citizens this can be accomplished in the writing of history and geography textbooks for children, youth, and adults. Maps can express a state's views of the world. How are areas of the world portrayed where there is disagreement on boundaries and labels of political spaces? The labeling of political spaces on land or coastal areas and the languages used on the maps convey a distinct view or jurisdiction of disputed territory. During the Cold War, colors used on maps helped to distort the size of enemy forces to convey impressions of a national threat.

Traditional ways of conveying images include official visits, speeches, and votes in international assemblies. A new vehicle for communicating "who one is in the world" is the Internet. Many countries, regardless of size, location, ideology, and economic standing, now have official Web pages. On these colorful, attractive, and heavily symbolic "pages" is information about the state, aimed at the increasingly electronically oriented sectors of society, be they government offices, businesses, universities, or individual households. States are seizing this electronic medium to present in capsule form information about a variety of subjects: a thumbnail sketch of their official history, business opportunities (usually real estate and tourism), unique cultural features, and linkages to sports, music, food and wine, museums, language instruction, and current events. These "pages" can hypothetically be accessed by anyone from anywhere (where there is a computer terminal) for multiple purposes (currency values, weather, pending legislation, investment, travel, or virtual entertainment). The packaging of information on a state's official Web site reveals the way it wishes to present itself and how it views the world. The emphasis on symbols, colors, playful icons, boxes of information, and varying typefaces is a different way of presenting one's place in the world than previously, where the emphasis was on words. Symbols and images (maps, still photos, satellite coverage, and web-site technology) are rivaling words in a state's presentation of itself. Those states and leaders who want to effectively influence public opinion and enhance their own positions recognize the advantage of seductively packaging issues with a heavy dose of symbols, color, cartoon-like characters, doctored photographs, and a few carefully chosen "buzz" words.

States not only can construct their own views but can reconstruct views of their residents, friends, or adversaries. States that have experienced lengthy periods of conflict and oppression may seek to present a different "image," which they hope will facilitate their entrance into favored economic or cultural alliances. The expansion of the European Union and NATO to include former communist countries in Eastern Europe, for example, is based on specific policies these states must adopt, including the promotion of political openness, the protection of ethnic minorities, the easing of travel restrictions, and the creation of opportunities for outside investment; all of these steps require a public relations element to demonstrate the validity of membership application. Other states may seek to repair an image that was tarnished by recent scandal or by an event that cast the government in an unfavorable political light. Switzerland is a case in point; in 1997 it was forced to respond to adverse publicity associated with reports that large Swiss banks held Nazi gold during World War II; Switzerland eventually agreed to a large monetary settlement with Jewish survivors.

In the twenty-first century centers of political and economic decision-making and influence will likely shift from an exclusive North (or developed world) axis centered on Europe and North America to one more centered on Asia. Nearly 60 percent of the world's population now resides in Asia, and that percentage increases with each decade. Emerging Asian powers have long histories that are not dominated by cultural ties to Europe and North America. They may challenge the dominant European and North American views of individual and group human rights as well as the North's concerns about terrorism, weapons control, and environmental issues. The South's agenda may place higher priorities on eradicating disease, providing low-cost housing, eliminating famine, and reducing illiteracy. The South may devise strategies for technology transfer, community cohesiveness, collective civic virtue, and political participation that are different from those developed by and diffused from the West and North. The strong roots of religion may become major elements influencing domestic and regional economies and policies. The results of these North/West and South/East political shifts and frictions are uncertain, but what is certain is the importance of communication in the "Who gets what where" question. Political scientist Samuel Huntington writing in 1993 addressed this "clash of civilizations" between those in the West and those in the East. The "fault lines" lie between those affected by European ideologies and those holding Asian philosophies. How these competing worldviews play out will be important for all states as will the role of information and communication in the twenty-first century.

Key Components in Geopolitical Information Worlds

Five components are important in a study of the role played by the state in geopolitical information and communication: producers, gatekeepers, silenced voices, alternate voices, and consumers. The main elements in the communication cycle are the producers and the consumers of information. Although this discussion focuses on state-controlled information, nongovernmental organizations (NGOs) are also active players in the geopolitical information worlds.

The Producers

The state, often seen as the key player in the information world, defines itself in words, symbols, documents, and advertisements as well as through maps. It designates the official language or languages, what is taught in the schools, what is used in broadcasting and the print media, and the discourse employed in establishing laws, ratifying treaties, and conducting scientific exchanges of information. The introduction of a standardized grammar, speech, and alphabet also affects the standardization of regional dialects. The state decides what information will be collected about individuals and households in censuses. Some information is produced by and for the state, some for other consumers. The same applies to atlases and maps as well as photos and news releases. Some of the information may concern how the state has solved a social problem or how it wants others inside or outside the state to see how a problem is being solved. In the political and military arena, the state also defines its friends and adversaries not only on maps but also by words and phrases. Laws, regulations, standards, and qualifications are all defined by the state. All levels of government produce information because it is demanded by different administrative layers but also because the state seeks to justify its own existence by what it produces.

The Gatekeepers

Gatekeepers are those who are responsible for what information is produced. The state may be considered a gatekeeper, but within the state, there are individuals, agencies, and bureaus that have the final say about what is said and how it is said or presented; editors and television station executives, for example, are gatekeepers. The biases of the gatekeepers are also reflected in what information is produced under their aegis. There are also government ministries of information and culture, whose information produced and disseminated may also be slanted. Information may be deliberately distorted in numerous ways—through biased views about an ethnic or racial minority, through underreporting of major diseases, and through withholding information on an environmental disaster. The maps reflect how the state looks at the world. Official government newspapers, broadcasts, and telecasts reveal what the state wants to say to its own citizenry and to foreigners. The state also influences what children study in schools via textbooks, and it funds film documentaries, museum and art exhibits, television programs, scientific research, and government publications. That information may be fair or biased.

The state is comprised of many sources, avenues, conduits, and channels of information. These send information to other government units within the state and to other states and nongovernmental organizations. The state also has "damage control" operators to correct wrong or misleading or unacceptable news reports from journalists and other sources. All these items reflect something about how the state looks at itself and what it wishes to convey to others about itself. The levels of gatekeeping can vary widely, from a policy that encourages open reporting (freedom of the press) and the presentation of information or one that is heavily censored. The failure of the Soviet leadership to broadcast a warning of the Chernobyl accident to its citizens in April 1986 was a factor contributing to the downfall of the Gorbachev government. Instead the detection and subsequent spread of the radiation cloud were reported by northern and central European states and scientists.

Examining the nature of information, the basic question is, Who speaks for the world today? Dominant speakers have been white, wealthy males, those who control by ownership and censorship, and influential lobbying groups with specific religious, ideological, constitutional, and corporate agendas. We are now witnessing a growing micropress and microcommunications industry for special interest groups that were formerly marginalized. As a result, advances in narrow (TV) casting, video production, desktop publishing, listservs on a host of topics with subscribers from many countries, and computer networking are giving voice to groups once silenced—women, African Americans, Hispanics, gays, the elderly, the disabled, and those disenfranchised and linguistic minorities.

The rise of these specialty presses and publications also reveals a growing unification of political groups formed around identity politics. Special interest magazines such as *Maturity* (for the elderly), *The Advocate* (for lesbians and gays), *Signs* (for feminists), and *Charisma* (for conservative Christians) represent not just sources of entertainment but also vehicles for the dissemination of information and political strategies. Notices of important pending legislation or news items chosen and phrased to enhance a group's solidarity and viewpoint relate to a state's political geography. Indeed, an examination of where these magazines' subscribers reside would reveal much about a region's political geography as would the e-mail addresses of those subscribing to listservs on various global topics.

Recent media history has focused on boycotts lobbied against certain television programs and their advertisers by groups opposing the content of these shows. And although theaters in conservative communities may have refused to show a controversial movie, this same movie can now be seen on video in many of these same communities. This raises an intriguing question: Does such a movie have less of an impact when it is viewed privately rather than publicly? MTV has also been the subject of debate concerning its inclusion in cable packages. Controversies over obscene lyrics highlight variations in prevailing social climates in different areas. Local Public Broadcasting Service (PBS) affiliates also differ in terms of which controversial programs they air, including programs dealing with AIDS, depictions of gay relationships, paramilitary units, and religion.

Controversies over certain books point both to the political ramifications of media in different parts of the world and the interconnectivity of cultures. The diverse political reactions to Salmon Rushdie's *Satanic Verses* by Muslim and Western powers reflected a contrast in political and cultural perspectives. And interestingly, although the British government condemned the bounty put on Rushdie's head by the Iranian government, the United Kingdom banned the book *Spycatcher* for its revelations of sensitive aspects of British espionage. Thus, whether it is television, shock radio, or a book, "entertainment" can become a means for socio-political controversy.

In the former Soviet bloc, dominant voices are still those of males, members of the leading ethnic groups, and government-sponsored media. The emergence of democracy in the former communist countries of Eastern Europe has lessened government reporting as the major and official voice, but communication bureaucracies remain powerful. However, women's networks are now demanding increased representation in Eastern and Central Europe's changing societies. Minority ethnic and nationalist movements also face difficulties in publishing in their native tongues. We can expect more voices to emerge in the coming decades, such as those of dissident groups and returning émigrés.

In most poor countries the voices speaking today are those of the wealthy ruling elites, the multinational corporations, and the imported press, videos, and television programming. Indigenous voices and those with less access to and control of telecommunications are often drowned out by First World communications.

In all parts of the world, the state response to technological innovation has been partly political, often with unintended results. Thus, regulation of the exploding networks of cyberspace created by electronic mail and the use of the Internet for entertainment and education is being debated. And only in the last few years have U.S. politicians begun to recognize the potential of electronic town meetings, e-mail links to their regional and Washington offices, and the dissemination of information on the Internet.

China's entrance into the international business arena was via the FAX machines in many universities and businesses. During the Tiananmen Square massacre, overseas contacts relayed foreign news reports to the mainland via FAX, effectively bypassing the official Chinese news blackout. These same FAX machines were used to get information out of China. The Zapatista rebels in the southern Mexican state of Chiapas have their own Web site, as do other revolutionary groups and those seeking self-determination. Today many governments in traditional societies and emerging democracies are weighing the impact of new technologies, foreign entertainment, and news media networks on the culture, values, and political structure of their countries. After "plugging into" First World technologies and programming, they are finding out that social influences from the Western media can influence everything from dietary preferences to morality.

Silenced Voices

Despite the advances in information and communications technology and the ongoing discussions about "global villages" and "instant" communications and news, some groups will still not be heard. Silent voices often are the victims of discrimination and inaction. They lack power because they cannot acquire the information and requisite technology needed to be a part of the worlds around them. These silent voices also exist in isolated and inaccessible rural areas.

Threats, censorship, and imprisonment were among the strategies used to silence individuals and groups in the former communist bloc. Silent groups still include dissenters (scientists, poets, philosophers, housewives, and so on), foreign voices (émigrés), ethnic minorities, those supporting religion, the aged, war veterans, AIDS victims, the poor and uneducated, and exploited women.

Politics can be viewed as the struggle for multiple voices to be heard, with dominant voices at times suppressing others. For instance, what percentage of the U.S. population has legally been eligible to vote throughout American history, and how has that varied regionally according to state laws? In college textbooks and historical atlases of the United States, how are Native Americans represented? A survey of several leading textbooks and atlases finds that settlement maps of North America almost never depict Native American–controlled areas as coexisting with European "settled areas," thereby helping to silence Native Americans' territorial claims.

Alternative Voices

"Alternative voices" are nontraditional views within a society. Usually there are only limited, if any, channels for input from alternative voices, so most often these voices are expressed through similarly nontraditional means, that is, other than voting or working with specific branches of government. Alternative voices tend to challenge the status quo of power as well as society's accepted views by introducing perspectives and values that may not have a niche within the current system. We can see examples of alternative voices at different scales of human activity. At the grassroots or local level, these voices may take the form of neighborhood groups protesting decisions made by the county or state on the site of a community landfill or a prison. Informal groups who do not accept the proposal may choose to post flyers, circulate a petition door to door, hold public demonstrations, or conduct other forms of communication to garner additional support in pressuring the decisionmakers.

At the national scale, alternative voices also express nontraditional views on how states should manage their citizens, resources, and relations with other states, and they are generally located toward extreme ends of the political spectrum. Alternative voices range from evangelical religious and paramilitary groups to feminist, human rights, and radical ecology organizations. Even at the international scale, alternative voices challenge current power structures. Some organizations speak on behalf of "silenced" or "persecuted voices." Amnesty International is an organization concerned about the human rights abuses of citizens within their own country and those held by others. Their annual report contains information obtained by groups that operate in fear of retaliation.

The Consumers

We are all consumers who receive information from various technologies, including telephones, radios, and televisions. Consumers are also identified by their occupation, education, income, and lifestyle. Some information sources are geared for specific groups, and others are for specific places and regions. Governments also target publications, public service announcements, and programs for specific clienteles.

Within Europe, for example, the number of telephones per capita varies widely, especially when comparing the former communist countries of Eastern Europe with the West. Consumers receive information about political events from the state and other information organs on a daily basis. Those who have access only to single sources or irregular amounts of information will have different worldviews than those who receive input from multiple information sources. The media have no political boundaries, and radio and television broadcasts opposed to governments can sometimes transcend state borders. Voice of America and Radio Free Europe during the Cold War, for example, provided residents of Eastern Europe with "official" news about Western Europe and about events within their own states.

The Emerging Electronic State

In this age of electronic media, satellite television and the Internet are increasingly connecting people with distant images and sources of information. What kinds of political implications accompany this trend? One provocative view comes from Joshua Meyrowitz who believes that the use of electronic media creates a gap between physical place and information access. He argues that using electronic media alters the boundaries of social behavior by eliminating the necessity of interacting within a particular place. The media not only provide us quicker or more thorough access to events and behaviors but also provide information about new behaviors. According to Meyrowitz, as electronically mediated social situations transcend physical place and as media enable greater homogeneity among users, "place" loses its meaning as a unique social construct. We might relate this view to global-reaching phenomena such as international trade and travel, which may be seen as overcoming the obstacle of distance as well as state boundaries.

The Internet allows one to gather information on an infinite number of topics from all over the world in a matter of minutes. Trademark symbols of McDonalds's restaurants, Coca-Cola, Pepsi-Cola, Pizza Hut, and other multinational corporations are recognized around the world due to expanded trade and manufacturing practices. Sit-com reruns from the United States have become a familiar import of industrializing and newly industrializing countries. Advertisement campaigns by American Express Travelers' Cheques and major hotel chains assure us that no matter how far away we travel, we can still enjoy familiar comforts and the conveniences of home. All of these images suggest that, indeed, state boundaries are permeable to information flows and that places are becoming more similar due to more homogeneous access to information. So has place lost its meaning?

Political geographers would argue "no." Place is still an important aspect of daily human life and of the corporate and political worlds. Tourism advertisements, for example, emphasize the unique, exotic, geographically and culturally specific characteristics of a place. Countries pay for large and expensive advertisements, usually with photos, maps, and the phone and fax numbers of banks, to attract foreign investors. States promote a particular image that emphasizes the individuality of place, including factors that help promote sex-based tourism. Images, whether in print or on the screen, are a major part of a state's boosterism. The return of Hong Kong to China after 99 years of British rule was accompanied by dance and music ceremonies that focused on Chinese tradition. This, in addition to the Chinese troops being moved in place for the transition, emphasized the extension of China's territorial borders both in symbol and in fact. Similarly, when Laurent Kabila gained control of Zaire in mid-1997, he changed the country's name to the Democratic Republic of Congo (this includes the country's previous name, "Congo"). This act projected a new image to distinguish his leadership and to invigorate the country's sagging economy. A new name suggests a new place, and using an old name in the new name suggests a return to past order, values, and stability within a democratic framework. Indigenous groups also recognize special places in their heritage, places that previously may have been destroyed physically or removed from memory by outsiders and imperial governments. A part of their spiritual and cultural geography may be embedded in sacred mountains and rivers, the sites of birthplaces of fallen heroines and heroes or victorious conquests. They wish to see these new landscapes of memory and nationalism fashioned in their new classroom geographies and histories as well as museums, anthems, holidays, and revived folklore. We also see many states, large and small, old and new, taking advantage of the Internet to promote a global identity with their unique place.

Other groups engaging in electronic communication seek to influence public policy via the mobilization of interested parties. These include those with local and grassroots interests and those with global agendas. Their objective is to utilize electronic media to influence policy decisions in favor of a given decision supporting their own interest. Women, gays, religious groups, and environmentalists of various shades of green are among those who mobilize their supporters to form public interest groups. By the beginning of the 1990s more than 2500 public interest groups had formed in the United States. Although some have substantial funding and resources at their disposal, many must compete for funds and support from foundations, corporations, and individuals in order to maintain their operation. Because this competition for monetary and supportive resources can be fierce, communication with potential donors is of utmost importance. One of the main non-electronic forms of communication relied upon by public interest groups is direct mail; it is used to educate the public about the group and the issue it is advocating.

Since the ultimate aim of public interest groups is to influence a policy decision, other means of communication are used to direct public demands to the relevant government body. Writing letters to one's congressional representatives, for example, has long been a means of communicating public opinion to decisionmakers. More recent forms of communication to achieve the same end rely on computer systems. The Clinton administration was the first to be "wired" to the Internet for the purpose of opening a channel for public input. Any citizen with access to e-mail can send a message to the president within a matter of minutes. Another computer-based means of influencing decisionmaking is operated by an organization called the Clinton Group (no connection to the Clinton administration). This organization works with interest groups by matching group rosters with individuals' telephone numbers. Using an automated telephoning system, the computer dials the number for each person. When someone answers, an operator explains the issue being advocated and then offers to connect the individual—free of charge—to the switchboard of his or her local congressman. In this way interest groups are able to deluge congressional representatives with input regarding the particular interests of that group. When controversial domestic (immigration, welfare reform, abortion) or international topics (conflicts with adversaries) are on the public agenda, the switchboards of representatives "light up" and the e-mail volume is heavy.

Information Landscapes

States contain a variety of information and communications landscapes. They include places that produce, consume, and manipulate information as well as store and access it. The features that one usually associates with the landscape are government offices for the executive, judiciary, and legislative branches. These may include a few or many buildings in a country, depending on its size, wealth, information economies, and networking with other states. Additional elements in that political information landscape are state-supported universities and colleges, banks, research laboratories on a wide variety of products (food, drugs, motor vehicles, computers, weapons, etc.), medical facilities, libraries, and offices for the print and visual media. Intelligence-gathering offices and ministries include diplomatic missions and various military attachés. Local citizens are connected to the national offices of government agencies and bureaus through a series of electronic networks, including phone, FAX, and computer systems. The state is integrated through a variety of information-gathering and disseminating networks, including post offices, schools, employment offices, and hospitals, as well as official radio and television broadcasting. Increasingly, official government offices list telephone and FAX numbers in addition to postal addresses for regional offices as well as embassies and consulates on their official Internet homepages.

Information Economies Receiving State Support

The postindustrial society, with its emphasis on information, is spawning a host of new industries and services. These electronic commerce worlds include entertainment and leisure, banking and finance, real estate, communications and telecommunications, environmental monitoring and regulation, education, and health. All these are knowledge industries in the sense that they depend on the production, consumption, and dissemination of information. Their evolution and diffusion throughout a society is closely tied to the state. The state, whether at the local or national level, is part and parcel of "what is provided for whom and where."

The state has become a major partner in the information economies that are assuming larger and larger shares of the workforce. Three fields where the business-state linkages are strong are health, environment, and entertainment. One might even argue that "health is information," in the context that diagnosing a health risk, settling the eligibility levels for those admitted to hospitals who are covered by medical programs, and payments to physicians treating patients are all involved in health care decisionmaking. The "environment" also conjures up many meanings related to information and politics, including legislation defining endangered and threatened plant and animal species, labeling (and mapping) wetlands and safe habitats for unique species, and setting safety and health standards. Entertainment is a growth industry or service, and much of that growth is often supported by state incentives, whether it be casino gambling and other forms of "commercial sins" or Hollywood films "popularizing" history. The sporting world's ties with government extend from local support to build new professional sports stadiums to approving new television sports-focused channels for regional audiences and sponsorship and training of team members for the World Cup and Olympics.

Information industries seek economic incentives when locating their services in new places, often overseas. They lobby heavily for regulations and licensing that will benefit and secure their position and make it difficult for newcomers to compete. They seek advantages in selling their products to emerging markets, whether videos, movies, computers, or computer software or intellectual property rights for their products. Information moguls are transnational and becoming more so with acquisitions of smaller companies. In examining the future geopolitical and information map, one needs to consider the growing influence of information and communication corporations such as Microsoft and IBM. The national and global networks of these new foreign policy actors extend from financial support for those seeking national office to those on global decision boards supporting entry into new markets.

Internet Worlds

Perhaps no single technological innovation has so shaped the past decade of the twentieth century as has the Internet. Evolving from partnerships between the American defense industry and largely university-based computer scientists, this technology was originally created as a defense communication system. Thus was born the infant Internet, a de-centralized system capable of interlinking multiple communication nodes that would be retained should other systems be destroyed or become inoperative. Electronic mail is the most familiar product of this development. The speed, efficiency, and eventual low cost of e-mail have subsequently been marketed to business and now the computer consumer. Further innovations have created databases and methods of storing for enabling multiple access, paving the way for the Internet's plethora of websites, on-line banking, entertainment, and similar virtual realities.

The population using the Internet is not only increasing, but it is also broadening. In 1994, 1,240 Internet servers (computers storing files for access via modems) were operating, and the vast majority (85 percent) of Internet users were male. Within three years there were more than one million servers and 44 percent of the users were female. By 1997 a host of economic activities found a ready haven on the Internet, such as on-line banking. Almost every major and many small companies now host a website with a range of product descriptions, customer service guides, and even on-line purchasing. Travelers can now use the Internet for tracking flights, fares, hotels, and auto rentals. Higher education, one of the birthplaces of the Internet, has not been immune to this frenzy; universities are now developing a complete on-line registration process. Once electronically registered, students can progress on-line from reading course descriptions to registering for classes. The student then receives a listing of her or his classes, a map of the class location, a required reading list, book prices, tuition billing, add/drop/refund dates, and even an on-line video of bookstore lines!

The anonymity of the Internet and the ephemeral nature of "chat rooms" that exist only as long as people are in them provide easy access to information on a host of taboo, controversial, and illegal activities ranging from marginalized sexualities and prostitution to espionage and censored materials. No longer must a person resort to dark alleys, whispered public phone conversations, or marginalized adult bookstores to discuss, view, sell, or order controversial ideas and information. Bootleg sound files, pirated software, banned novels, censored political writings, and the burgeoning on-line sex trade are creating legal and political situations that reveal the spatial qualities undergirding off-line and socio-legal institutions.

Innovations in hardware and software pose problems for the state, including the development and enforcement of laws to protect users and hardware and software producers. Laws are defined like the state itself, by a fixed spatial jurisdiction; what is a crime in one jurisdiction may be a government-regulated business in another. In communications law, court decisions have focused usually on several key elements. In some cases, countries may prohibit and punish the end user of a product, such as someone who reads censored materials. In other cases, the law applies punitive or licensing measures on the producer of an image or product tied to the location or access of the product's source. In still other cases, such as copyright infringement, the actual material is not itself illegal to consume as far as content, but rather the nature of the dissemination is banned.

Previously, communication technologies such as newspapers, radio/TV stations, and publications usually came from a single point source. Regulation of such material then relied on control of the point of product or access points emanating from this source into a given jurisdiction or spatial unit. The Internet, by contrast, is multinodal with an average user able to transmit information to potentially millions of end users without the infrastructure tied to a radio/television station or a printing press. The relaying of this information by other users to potentially millions more produces a scenario analogous to being able to produce a news story that is transmitted simultaneously to hundreds of television stations. In an attempt to regulate on-line content, the U.S. government in 1995 passed the Communications Decency Act as part of a much larger telecommunications de-regulation bill. A 1997 unanimous decision by the U.S. Supreme Court ruled that the government could not impose mandatory labeling, filtering, or other censorship devices on Internet content, thereby giving on-line speech the constitutional protections of free speech afforded to off-line print media.

Commercial researchers in digital versatile discs (DVDs), the next expected innovative electronic product that combines CD and VCR recording/viewing capabilities, have designed six different and noncompatible formats. Each format is compatible to a particular region of the world but cannot be played or copied in the other five. China, with its reputation for widespread copyright violation and recording piracy, has its own format as one of the six regions. In this way, companies hope to address copyright violations and intellectual property law in the electronic era.

The regulatory future of the Internet thus may not be in traditional legal approaches but more in line with laissez-faire market forces, technological structural safeguards, and campaigns urging ethical self-regulation over a medium whose innovation pace and design largely hamper traditional approaches to geopolitical controls over information.

Visualization of Foreign Policy

The introduction of television and satellite imagery in reporting events around the world, regardless of location, plus the recognition that speed is an element in relaying information on political changes, social conflict, and environmental catastrophes, has brought into light how these factors can affect a country's foreign policy.

Visual images of people, places, and situations are significant ways in which governments and citizens learn about events and problems. Many government leaders as well as citizens will state that their knowledge about an issue or personality came from "the evening television news" or the photographs that appeared in newspapers or magazines. The "CNN effect" can include the very moving images of refugees fleeing burned and bombed villages or rebels assuming control of government buildings.

The camera is an increasingly important device determining "what people know about what goes on where" in the world. Some governments may seek to ban or curtail the visual coverage of cultural and economic problems within their own boundaries, but that objective is difficult to fulfill by licensing photographers and when citizens are able to transmit their visual messages across borders. Even in states where the reporting of events is an accepted part of the freedom of expression, there is a concern that private broadcasting media may distort or misrepresent the state's views. The government in such cases may seek to blame the media for exploiting a situation or unfairly representing a state's actions or viewpoint. But states themselves are very powerful sources of visual information.

Greater visualization of a country's foreign policy concern can have several impacts. One result may be that the state will become more dependent on photographs and satellite images to educate a public about its views. Another is that words may become a secondary means of informing the public about a crisis. A third is that the opportunity to mislead the public may be easier to accomplish through pictures than words. That deception may occur with doctored photographs, seductive colors used on maps, or increased symbology. Each of these media could be used by a state or a powerful media corporation to fit its own political agenda.

Information "Wars"

Wars are fought with and over information, which explains why revolutionaries seize as initial targets broadcasting and publishing facilities. Fu-

ture wars may even arise more from information conflicts than from traditional disputes over land or water. Information "wars" may target intellectual property (who owns what and whose rights have been violated and at
what costs), expansion of bandwidths for use in broadcasting, and the
spread of viruses to destroy an opponent's ability to communicate. International information disputes are likely also to emerge in the labeling of products, consumer product safety, favorable and unfavorable trade practices,
measures to halt the spread of diseases, and the penalties meted out to
those who are environmental polluters, destroyers of culture, disseminators
of violence, and denyers of basic human rights.

Geopolitical Information in 1900, 2000, and 2100

Geopolitical information and communication varies over time as well as
space. Here, we explore the geopolitical information worlds in 1900, 2000,
and 2100 (see Table 16.1). We selected these three centennials so that the
mixes would be different; over time, technological advances in communication and information processing influence what is produced, how it is communicated, who the consumers are, and who the gatekeepers are. By 2100,
perhaps the state itself will be replaced by another political structure, but
information will continue to play a major role in what decisions are made,
what problems are solved, and what data are mapped.

The Worlds in 1900

For purposes of this discussion, the world at the start of the twentieth century can be considered as a rich world–poor world dichotomy with a European–North Atlantic "core" and a colonized periphery. We define the core
as Europe and North America, the periphery is everywhere else. The information and communications worlds of the core and the periphery were
quite distinctly different.

The core communicated via the emerging modern communication systems, while much of the poor world still used the oral tradition. The information networks in 1900 included limited telephone usage, weak international connections, slow mail, and intermittent wire services. The few
major geopolitical centers were located in the core. Geographic information produced in 1900 concerned land use, rivers, agricultural production,
minerals, transportation networks, and new industries. Key geopolitical information included that on colonial holdings, European power shifts, and
the rise and fall of leaders. Information of this sort was presented in official
pronouncements, documents, reports, statistics, and maps. The key gatekeepers of geopolitical information were the major powers and the information "houses" of the major industrial centers. The gatekeepers were the

elite, rich, powerful, white males who resided in the centers of political and economic power.

The Worlds in 2000

The worlds at the end of the twentieth century will reflect more geographic variation than at the turn of the previous century. Although the rich world has computers, FAXs, satellite imagery, multiple television channels, and multiple newspapers and radios, these sources and services are less available elsewhere. In the poor periphery, there is still a reliance on radios, newspapers, limited television, and an irregular postal and telephone system. The widespread usage of advanced communications technologies in the core and the lesser rates in the semiperipheries make up one of the realities of the late twentieth century, mainly as a result of economic disparities.

Information at the end of the 1900s will be conveyed via satellite, videocassette recorders (VCRs), and electronic bulletin boards, with multiple centers producing and generating information. Geographical information sought in 2000 will focus on environmental damage, movements of capital, UN voting, and activities of NGOs. Geopolitical information will be sought regarding global famine forecasting, global grain and oil production and prices, elections in new democracies, satellite coverage, and transborder datasets. The gatekeepers as we approach the twenty-first century will be the state and groups that are interested in what goes on within and beyond the state. These include satellite and computer producers, media and public opinion corporations, research and development labs, and libraries. Key concerns will include what information will be admitted by the gatekeepers, how it will be disseminated, and how it will relate to the goals of states, NGOs, and the formerly silent voices.

The Worlds in 2100

Information demands some 100 years from now may be less about traditional geopolitical divisions and more about the earth, satellites, and planets. In what we might call the "expanding universe," Earth will have launched and established permanent satellites as well as colonies and settlements elsewhere in the universe. Information may encompass fourth- and perhaps fifth-dimensional worlds. We will see further advances in video phones, virtual reality, and portable, instantaneous global communication. Extrasensory perception may be feasible, replacing the mysticism and astrology of previous centuries. Populations are likely to be connected by transglobal phone directories, postal and money systems, and instant online services among major decision centers scattered across the planet and beyond. Examples of geopolitical information might include the growing

TABLE 16.1 Elements of the Geopolitics of Information and Communication, 1900–2100

Forms of Communication (Technological Innovations)	Who Speaks and Has Power (Gatekeepers and Producers)	Silent Voices (The Powerless)	What Is Important and Mapped (Information Regions)
1900: Rich World			
printed word	Eurocentric world	women	two-dimensional world
infant telephone and cinema	the rich & powerful white, Protestant,	exploited labor new immigrants	agricultural productiv- ity
limited wire service	wealthy males	nonunionized labor	terrain & land uses
photography	colonial powers	illiterates (including	transport networks
newspapers	large companies	women)	European power shifts
infant radio	unstandardized	racial/other minorities	resource sites
postal service (nodal)	national presses		military targets
word of mouth	political elites		colonial holdings
art	few correspondents/		time zones
currency	news wire services		
1900: Poor World			
word of mouth	white colonists	indigenous masses	resources, tribal areas,
oral traditions	white missionaries	exploited masses	and transport
art, dance, painting	white commercial elites	native voices	networks involved
postal service (nodal and usually colonial)	some assimilated native speakers	suppressed cultures	with colonial administration
intuition: astrology, tarot, mysticism			and extraction
personal couriers			
2000: First World			
printed word	editors	the poor	environmental
radio	writers in major	the uneducated	inventories
television and VCR	languages	unorganized minorities	biological diversity
postal service (home delivery)	radio/TV producers large NGO leaders	minority language writers/speakers	resource reserves standardized regions
credit reporting	wealthy white males	victims of bigotry	(bar codes, ISBN,
dominance of books/TV	neocolonial powers televangelists	access to info nets computer illiterates	etc.) minority group spaces
computers	major universities	rural areas	votes
standardization	software producers		political and legal
desktop publishing	special interest groups		cultures
FAX	more female/minority		influence of NGOs
electronic mail (instant node)	voices state leaders		TV reception areas newspaper circulation
newspapers	token minority voices		areas
psychological testing	media		creditor/debtor
statistical forecasting	large banks, creditors		relations
telephones/cellular phones	economists		distribution of electronic networks
			distribution of bank- ing/ATM networks
			cultural refuge areas
2000: Former Second World			
radio	dominant ethnic groups	dissidents	government-controlled
television	males	ethnic minorities	areas
selected/restricted FAX and e-mail	government media censors	women victims of bigotry	rebel areas "flash points";
telephone	selected dissidents	foreigners	politically, culturally
newspapers		religious minotiries	& ecologically
			satellite coverage
			weather forecasting

(continues)

TABLE 16.1 *(continued)*

Forms of Communication (Technological Innovations)	Who Speaks and Has Power (Gatekeepers and Producers)	Silent Voices (The Powerless)	What Is Important and Mapped (Information Regions)
2000: Third World			
radio	neocolonial powers	rural areas	religious influence on
imported media	First World companies	women	politics
pulpit	selected elites	classless masses	intellectual property
oral traditions	state-dominated media	rebels	rights
selected postal and	imported media	the poor	refugee movements
telephone networks		ethnic minorities	sexual slavery
			illicit drug trade
			human rights abuses
2000: Fourth World			
oral traditions	neocolonial powers	middle classes	
intuition	few rich elite	landless/classless	
popular leaders/pulpit	elderly and male	illiterates	
sparse phone/mail	traditional tribal voices	the poor (majority)	
2100: Earth			
telecommuting	multi-voiced world	uneducated	areas dominated by
electronic meetings	special interests	the poor	special interests
instant video access	NGOs	computer illiterates	corporate states
hypermedia	human welfare and	the inaccessible and	global/personal liberties
HDTV	environmental groups	isolated	information resources
e-mail (home)	technocrats		psychological profiles
portable phones			of world leaders
enhanced intuition and			World Court decisions
psychology in marketing			circumpolar regions
and the media			oceans, sea floors
plastic currency			intellectual property
uniform postal service			rights
cyberspace			trading blocs
multisensory imaginary			on-line global phone
and tactile systems			directories
replace drugs and			political parapsychology
prostitution (video			global ecological fore-
addicts)			casting
on-line democracy			cultural refuges
via electronic mail			genetic resources
fiber optics			
2100: Satellites and Space Stations			
no long distance charges	rich and powerful	poor and marginalized	planetary surfaces
global reach	intellectual elites		satellite environments
24-hour stock exchanges	corporations and states		communication grids,
standardization			solar winds, and their
			interactions
2100: Planets			
laser-coded messaging	nations in space	poor countries without	terraforming sites
interplanetary signal		space programs or	adaptions to new worlds
booster satelliltes		connections	
2100: Beyond . . .			
advanced bioengineering			interstellar/galactic space
systems carry human			
gene banks beyond			
solar space			

political power of the periphery, environmental changes, and the redistribution of planetary resources and reserves. We may also see further research on manipulating public opinion and determining the psychological behavior of leaders. More gatekeepers will be non-European, and they will include more international organizations and regional political councils. In addition, there will be many more kinds of spatial data collected, organized, and disseminated in 2100 for investigating planetary problems.

Ongoing Information Problems

Some common information problems are shared by the worlds of 1900, 2000, and 2100. Issues include the following:

1. With the wide array of information available, how do decision-makers identify reliable sources of information, especially in areas of transition? Also, what sources of information will be relied on most by citizens and voters in different places?
2. Can the state satisfy the information demands of its citizens and ensure equity in the information? This question will be relevant for both small, homogeneous cultural states and multi-ethnic states. What are the vehicles used to ensure equity in the information?
3. To what degree does visible information from the sciences influence state policy (e.g., El Niño, global warming, and pending energy crises?
4. What will be the role of NGOs in providing information used by the state?
5. Who will be the producers and gatekeepers of vital information? What will their intentions and biases be?
6. Are the emerging information worlds reducing the gaps between the rich and poor, powerful and powerless, cities and rural areas, democracies and tyrannies? Will there be an "information gap" between those who have access to computer-based information and those who do not?
7. How will the states and other information producers use maps and spatial datasets to solve problems?
8. Where will this "blizzard" of information lead the state and the individual? Who will decide what will and will not be published and disseminated? Who will decide where the information will be available and at what cost? And who will decide what will be discarded?
9. Relating to privacy, what rights will a citizen have in a government that collects more and more information about individuals?
10. What communication channels will silenced voices pursue most effectively, and what kinds of filters will inhibit participation in deci-

sionmaking? Will the silent voices be given an opportunity to be heard? What responsibility will the state assume in permitting these disenfranchised voices to produce their own information and to influence decisions made by the state? Women and racial and ethnic minorities are among those rapidly acquiring the skills to influence state policies.

11. To what extent will Americanization be a major global information process?

12. Finally, will the world move toward more uniformity or more diversity in the information that is produced and consumed?

Summary: Looking Ahead

We have sketched some promising linkages among political geographic aspects of information and communication. These concepts challenge our thinking about the role of the state and international organizations. Although the state has traditionally dominated the "information business," the growth of information industries and the introduction and rapid proliferation of new communications technologies will redefine its role. Political geographic research examining information in a historical, contemporary, and futuristic context can contribute much to the debate about the role of information and communication in societal transformation.

References

Abler, R. "Hardware, Software, and Brainware: Mapping and Understanding Telecommunications Technologies." In *Collapsing Space and Time: Geographic Aspects of Communications and Information,* ed. S. D. Brunn and T. R. Leinbach, pp. 31–48. New York and London: HarperCollins Academic, 1991.

Abler, R., et al., eds. *Human Geography in a Shrinking World.* North Scituate, MA: Duxbury Press, 1975.

Adams, P. "Protest and the Scale Politics of Transformation." *Political Geography* 15, no. 5 (1996): 419–444.

Anderson, B. *Imagined Communities: Reflections on the Origin and Spread of Nationalism.* New York: Verso, 1989.

Annis, S. "Evolving Connectedness Among Environmental Groups and Grassroots Organizations in Protected Areas of Central America." *World Development* 20 (1992): 587–595.

Barnes, T. J., and J. S. Duncan, eds. *Writing Worlds: Discourse, Text, and Metaphor in the Landscape of Representation.* New York: Routledge, 1992.

Brunn, S. D., J. Jones, and D. Purcell. "Ethnic Communities in the Evolving 'Electronic' State: Cyberplaces in Cyberspace." In *Political Boundaries and Coexistence,* ed. W. Gallusser et al., pp. 415–424. New York and Vienna: Peter Lang, 1994.

Brunn, S. D., and T. R. Leinbach, eds. *Collapsing Space and Time: Geographic Aspects of Information and Communication.* New York: Harper/Collins and Routledge, 1991.

Foucault, M. *The Archaeology of Knowledge.* Translated by A. Sheridan Smith. New York: Pantheon Books, 1972.

Huntington. S. P. *The Clash of Civilizations: The Remaking of World Order.* New York: Simon and Schuster, 1996.

Kellerman, A. *Telecommunications and Geography.* New York: Halstead Press, 1993.

Kirsh, S. "The Incredible Shrinking World: Technology and the Production of Space." *Environment and Planning D: Society and Space* 13, no. 5 (1995): 529–555.

Luke, T. "Placing Power/Siting Space: The Politics of Global and Local in the New World Order." *Environment and Planning D: Society and Space* 12 (1994): 613–628.

Mackenzie, J., ed. *Imperialism and Popular Culture.* Manchester: University of Manchester Press, 1994.

Meyrowitz, J. *No Sense of Place: The Impact of Electronic Media on Social Behavior.* New York: Oxford University Press, 1985.

Neuman, J. *Lights, Camera, War: Is Media Technology Driving International Politics?* New York: St. Martin's Press, 1997.

Nierop, T. *Systems and Regions in Global Politics. An Empirical Study of Diplomacy, International Organizations, and Trade, 1950–1991.* Chichester and New York: Wiley, 1994.

Smith, D. A. *Human Geography: A Welfare Approach.* New York and London: St. Martin's Press, 1977.

About the Editors
and Contributors

John Agnew is professor of geography at the University of California at Los Angeles.

Stanley D. Brunn is professor of geography at the University of Kentucky.

Saul B. Cohen is professor of geography at Hunter College, New York.

Susan L. Cutter is professor of geography at the University of South Carolina.

George J. Demko is professor of geography at Dartmouth College.

Colin Flint is assistant professor of international affairs at Georgia Institute of Technology.

Alan K. Henrikson is professor of history at Tufts University.

Jeffrey A. Jones is a Ph.D. candidate at the University of Kentucky.

David B. Knight is professor of geography and dean at the University of Guelph, Ontario, Canada.

Phyllis Mofson is senior management analyst for sustainable development, Community Affairs Office, State of Florida.

Richard Morrill is professor of geography at the University of Washington.

Alexander B. Murphy is professor of geography at the University of Oregon.

Shannon O'Lear is assistant professor of geography at Illinois State University at Normal.

Marie D. Price is associate professor of geography at George Washington University.

Bradford L. Thomas is a private consultant and professorial lecturer, Department of Geography, George Washington University.

William B. Wood is director of the Office of the Geographer and Global Issues at the U.S. Department of State.

Index